Cloning

A Biologist Reports

Cloning

A Biologist Reports

Robert Gilmore McKinnell

Professor of Genetics and Cell Biology
College of Biological Sciences
University of Minnesota

UNIVERSITY OF MINNESOTA PRESS □ MINNEAPOLIS

Published by the University of Minnesota Press,
2037 University Avenue Southeast,
Minneapolis, Minnesota 55455
Printed in the United States of America
at Harrison & Smith-Lund Press

Library of Congress Cataloging in Publication Data

McKinnell, Robert Gilmore.
Cloning.
Bibliography: p.
Includes index.
1. Cloning. 2. Cell nuclei—Transplantation.
3. Embryology, Experimental. I. Title.
QH442.2.M32 596'.01'6 79-10569
ISBN 0-8166-0883-0

This book is for

the children

of

Beverly Walton Kerr

Preface

I wrote this book. But, it would never have appeared had I not received help from teachers and others over the years. Let me acknowledge their help, while not suggesting that they are at all responsible for whatever errors, in fact or in judgment, may occur in this book.

First, the teachers. Professor Lora Bond of Drury College inspired me, Professor Richard A. Goff of the University of Oklahoma encouraged me, Professor Nelson T. Spratt, Jr., of the University of Minnesota helped me, and Dr. Thomas J. King of the National Cancer Institute, instructed me.

Others assisted with the manuscript and its several versions. Most of all, I thank Ms. Beverly Kaemmer of the University of Minnesota Press for her friendly attention to detail and her gentle ways as she scissored and rearranged awkward prose. Dr. Barry Pierce, University of Colorado Medical Center, Dr Marie A. DiBerardino, Medical College of Pennsylvania, and Dr. V. Elving Anderson, University of Minnesota, each read the entire manuscript. Drs. Peter D. Ascher and Donald B. Lawrence of the University of Minnesota advised me about plants. I thank Dr. Edmund F. Graham, University of Minnesota, for discussing with me agricultural applications of cloning. My associates in research, Mr. Lyle Steven, Ms. Janet Sauer, Mr. Thomas Fontaine, Jr., and Dr. Robert Bergad read proof, detected errors, and made other suggestions. I thank them.

The frogs described in this book are a major part of my life. I thank Mrs. Ruth S. Boyce, Alburg, Vermont, for providing quality frogs for my research from the Lake Champlain area of northern Vermont. Mr. Gib Hedstrom, Alexandria, Minnesota, provided extremely useful advice and helped me with frog collections many times. My wife, Beverly, and children, Nancy, Robert, and Susan, have given up many a weekend, without compensation, to help collect *Rana pipiens* for my research. Their expertise in frog collecting and their sympathetic understanding are gratefully appreciated.

R. G. M.

1 April 1979

Contents

Cloning

A Biologist Reports

1

Why a Discourse on Cloning? Of Apples, Frogs, and Humans

Cloning is much in the news. The public has been bombarded with newspaper articles, magazine stories, books, television shows, and movies—as well as cartoons. Unfortunately, much of this information is incorrect. Inaccurate information plus an understandable public concern about whether a human has been or will be cloned, with all the ethical and moral questions that raises, have resulted in a very distorted view of what cloning is and why biologists choose to clone. The real story may seem less dramatic, but in a way it is more heartening. It is an account not of the production of carbon-copy dictators, millionnaires, and Einsteins but of research that may provide answers to the very human problems of cancer and aging.

Let me provide several examples of inaccurate information and comment on the potential harm of such information.

A seemingly trivial—but, in fact, critical—biological flaw in an ethical discourse was published by Paul Ramsey in 1970. He asserted that cloning is "reproduction by enucleating and renucleating an egg that has already been launched into life by ordinary bi-sexual reproduction. The question, Shall we clone a man? means, Shall we renucleate the human fertilized egg cell?"

Ramsey, a Professor of Religion at Princeton University, described the prototypic cloning experiment of Briggs and King at Philadelphia as transplanting nuclei into "freshly fertilized egg cells" (*Fabricated Man: The Ethics of Genetic Control*, New Haven: Yale University Press, 1970, pp. 64-66). Actually, frog cloning—presumably the model system for human cloning—involves the enucleation of an *unfertilized* egg. And that makes all the difference. Why? First, there is a technical reason. In frog (and other amphibian) eggs, the maternal hereditary material of an unfertilized egg is physically accessible and therefore enucleatable. Not so the nucleus of a fertilized egg. No one, to my knowledge, has ever successfully enucleated a fertilized frog egg. If it *were* possible, I wouldn't be concerned. With human eggs, I would. And this is the second reason why it makes all the difference. Although I might consider examining, studying, manipulating, and dissecting a human egg untouched by human sperm, I would, in fact, be loath to contrive an experiment on an egg already "launched into life by ordinary bi-sexual reproduction." Such an egg could be obtained by only the most heroic means (fertilization occurs in the Fallopian tubes). Removal of a fertilized egg from the reproductive tract of a woman could be considered an abortion. To abort for purely experimental reasons is clearly unthinkable. Fertilization can occur in laboratory glassware. But Ramsey was not referring to test-tube, in vitro, fertilization, for surely this is not "ordinary bi-sexual reproduction." It seems to me that misconceptions of this nature engender apprehensiveness about contemporary biology that is not warranted.

In 1974 Joseph Fletcher, theologian and Professor of Medical Ethics at the University of Virginia School of Medicine, described cloning in essentially the same terms: "A fertilized ovum or zygote is extracted from the oviduct and the fertilizing done *in vitro*. Next, its nucleus is removed (enucleated) and a body or 'somatic' cell is donated" (*The Ethics of Genetic Control*, p. 71).

Recently, David Rorvik, a science writer, asserted that a wealthy businessman had had himself cloned. Shortly after Rorvik made his assertion, three scientists sought an investigation of and controls over cloning and related genetic research. Biologists who work with cloning in frogs and with related procedures that may result in mammalian cloning believe that Rorvik is incorrect. Their belief that the claim is erroneous is based on a quarter of a century of experimentation in cloning and manipulative reproductive biology.

A Congressional Subcommittee on Health and the Environment responded to the public's interest in this claim of human cloning with hearings in Washington, D.C., in May 1978. The subcommittee sought to ascertain if the claim was true and to look into the direction that research in cell and genetic biology is taking. Questions: Can a human be cloned now or in the future? Why did biologists undertake experiments that led to the uncritical acceptance by many of the unsupported claim that a human had been cloned? Several biologists testified before the subcommittee, but Mr. Rorvik twice failed to appear—the first time citing health reasons, the second an extended promotional tour for his book.

It would be exceptionally difficult to summarize the congressional hearings because of the number of questions, the complexity of those questions, and the many points of view that were represented. I think the testimony of Andre Hellegers of the Kennedy Institute for the Study of Human Reproduction and Bioethics, Georgetown University, reflects the attitude of many working biologists: "I think fundamentally the problem is that too many people believe that cloning is an end, namely, to production of an individual. Factually, cloning is a means. It is a means of cell study and an enormously important one" (*Hearings*, p. 90).

Public apprehensiveness about cloning could lead to reduced funding for cloning research or even to regulations outlawing the procedure. This could well impede advances toward a solution of such major medical problems as cancer and aging. Certainly hazardous procedures should be subject to appropriate constraints. But I hope this book allays apprehensiveness and convinces readers that the nuclear transplantation procedure known as cloning is not hazardous.

WHAT IS CLONING?

The word "clone," derived from the Greek "klōn," meaning twig or slip, refers to asexual reproduction, also known as vegetative reproduction. A few examples: house plants are easily propagated (cloned) from a twig or a slip; gardeners have been cloning potatoes for years. The edible part of the potato is an expanded stem known as a tuber, which, like other stems, has a number of buds or eyes. When placed in soil, each bud is capable of yielding an entire plant, and the crop so produced is a clone.

CLONING PLANTS

Plants usually reproduce sexually. The essential fact of sex in plants, or in animals for that matter, is that hereditary material from *two* individuals is joined to form a new creature. Just as each organism varies in a diversity of subtle ways from its fellows of the same species, so too the sex cells provided by that unique plant or animal differ. It is not difficult to understand why each oak tree, or each rose blossom, is unique when one comprehends the nature of sex.

Pollen grains, which make some people sneeze in the summer, grow and form tubes when they find the sticky female portion of another flower. As the pollen grows (germinates) into the female part of the plant, two sperm nuclei move down the pollen tubes. One of the sperm nuclei fertilizes the egg and a new generation begins. The details of sex vary with different plants. The basic point is that single-celled plants, sea weeds, fungi, mosses, and all higher plants, including ferns, evergreens, and flowering plants, have the capability of reproducing sexually. Because sexual reproduction involves the union of disparate sex cells, diversity is guaranteed.

Sexual reproduction is not involved in the cloning process. The cloned new plant does not result from the union of pollen-derived sperm with the ovum in a flower. The plant produced by cloning is a manifestation of the capacity for new growth of the old plant body. Since no sexual reproduction is involved in propagation with a shoot or a twig, the new plant is genetically identical to the old plant. Cloning is used in agriculture to produce high-quality, uniform products. Desirable apple varieties are grown by grafting the variety onto an ordinary host tree. Seeds that result from sexual reproduction of the palatable variety of apple yield plants with the expected heterogeneity and fruit that is quite variable. Since non-uniform apples are not sought at the market, apples are cloned.

Apples and potatoes are not unique among vegetatively cultivated crops. Named varieties of grapes, edible varieties of bananas, sweet potatoes, sugar cane, pineapple, asparagus, and many other agriculturally important plants—even garlic—are "cloned." The cultivation of some of these plants has been known for at least 4,000 years.

Horticultural scientists are now developing culture methods for small fragments of economically important plants, but the biological principle is the same. Cloning is asexual reproduction whether it involves the growth of a relatively large twig or slip of a plant composed of many millions of cells, the growth of a small cluster of cells, or even the growth of a single cell.

Professor F. C. Steward and his associates at Cornell University in Ithaca, New York, produced carrots from the progeny of single cells and small cell clumps. Steward placed carrot cells from a mature plant on a culture plate containing nutrients and a gelatinlike material known as agar. A disorganized population of cells grew on the nutrient agar plate. Isolated free cells were removed from the mass and cultured in a liquid medium containing coconut milk, other nutrients, and plant hormones. Embryo-like structures formed after repeated cell division in the liquid growth environment. Thousands of these structures developed roots and shoots and could be grown to mature carrot plants. What was accomplished with carrots was also effected in a number of other plant species including water parsnip, coriander, tobacco, and orchids. The vegetative propagation of single plant cells in the laboratory to produce carrots and parsnips is the ultimate in plant cloning—at least with regard to the size of the initial slip, which is, of course, microscopic.

Perhaps an even more surprising result in plant cloning was reported by Professor Armin Braun of the Rockefeller University of New York City. A cell from a tobacco plant *tumor* was grown and progeny of the malignant cell produced a normal plant. Braun's experiment involved shaking a plant tumor rapidly in culture so that single cells were obtained. A single cell was then cultured until it formed a mass of tumor cells. Fragments of the mass were grafted to healthy plants. After repeated grafting, the tumor cells became progressively more normal in appearance until stems, leaves, and flowers formed. The flowers produced seed which, when sown, produced normal tobacco plants. The importance of Braun's experiment in this discourse on cloning is that in no way can a single tumor cell of a plant be considered a sex cell. Braun's experiment is important because it was an early demonstration of the reversibility of the malignant state. Thus, the cloning of single plant cells, both normal and tumorous, in the laboratory may be grouped with other methods of vegetative propagation.

CLONING ANIMALS

Ordinarily, animals, like plants, reproduce sexually. However, there is a procedure that permits asexual reproduction in amphibians—toads, frogs, and salamanders. This procedure is nuclear transplantation and is widely referred to as cloning. Cloning might as well be referred to as vegetative reproduction of animals. However, I suspect it would be puzzling to refer to a frog as having been produced by *vegetative* reproduction.

There are several reasons why frogs were the first multicellular animals cloned. Frogs have an abundant supply of eggs and sperm that can be used by experimenters. A biologist may obtain from a single ovulation only 20 eggs from a mouse, but 2,000 from a frog. The fertilization and embryonic development of a frog, which ordinarily occurs outside the animal's body in ponds, is accomplished easily in the laboratory in glass dishes (in vitro). This permits direct observation of and experimentation on all stages of development (in contrast to the fetal development of a mouse, which is hidden because the fetus develops in the uterus). The frog embryo matures into an organism with brain, eyes, liver, and other organs not unlike those of humans. The late nineteenth-century and early twentieth-century experiments that provided the scientific groundwork for cloning were done with frogs and other amphibians. Thus, history, a relevance to humans, as well as abundance of eggs and sperm and ease of handling embryos in the laboratory, were probably the principal reasons why frogs were cloned first.

Cloning frogs is not new. Like plant cloning, the microsurgical procedure of placing a frog nucleus into an egg deprived of the hereditary material has been available to the scientific community for many years. Frogs have been asexually produced in some of the nuclear-insertion experiments. A cloned frog reveals that the donor nucleus had all the hereditary material necessary for complete development. Perfection and use of the frog cloning procedure depended on a sophisticated understanding of amphibian reproductive biology and an equally sophisticated skill in the manufacture and use of microscopic surgical instruments.

Successful nuclear transplantation in amphibians requires that the egg be enucleated, thus removing the maternal hereditary material contained in the egg nucleus. Then other hereditary material, contained in the nucleus from a body cell, is placed in the enucleated egg. The resulting cloned individual is parentless.

Biologically, a mother is a mother by virtue of the fact that she contributes hereditary material via the chromosomes of an egg. In cloning, the chromosomes of the egg have been enucleated, so there is no female parent. A father is a biological father by virtue of the fact that he has contributed hereditary material via his sperm. Since no sperm has participated in the development of the cloned individual, there is no male parent. Without a male or female parent, a cloned animal is a product of asexual reproduction. A frog produced by nuclear insertion is the exact analog of a stalk of commercially grown sugar cane—or a grafted apple. The frog, the sugar cane, and the apple are all produced by asexual generation.

Why produce a frog asexually? Why climb Mount Everest? Mount Everest, it is said, is climbed "because it is there." Most scientists just don't work that way. They seek answers to substantial biological questions when they do a cloning experiment. It would be frivolous indeed to dissipate precious research funds to clone "because it is there." Frogs are not cloned to produce new frogs. It is certainly a more economic use of resources and time to let frogs reproduce sexually, as they have been doing so well for millennia. Scientists use the cloning procedure to gain insight into biological phenomena such as differentiation, cancer, immunobiology, and aging.

Is the hereditary material of an embryo or an adult cell equivalent to that of a fertilized egg? Is the hereditary material of a cancer cell or an aging cell the same as that of a fertilized egg? More crucial is whether the hereditary material of embryos, adults, cancer cells, and aging cells can be manipulated, coaxed, or provoked into expressing hereditary potentialities similar or equivalent to that of the fertilized egg. More simply stated, is control of some of the most fundamental aspects of cell biology possible? "Why clone a frog" may be rephrased as "Why try to understand differentiation?" Cancer and aging are the most obvious areas where new understanding is needed. Many, including me, think that cloning may provide new and useful insight into these critically important biological problems. And cloning may even be helpful in overcoming rejection in organ transplantation.

Frogs were cloned sometime ago. What about human cloning? Is it apt to happen? Probably not soon—if ever. Why the difference? The difference is related to the nature of biological research. There is no need to clone humans in order to provide answers to the questions that are central to cloning research. As I said, frogs are not cloned to produce new frogs. However, human cloning might relate primarily to producing more people. Reproduction by cloning is an inappropriate means to reproduce more frogs or humans because species survive through genetic heterogeneity. Sexual reproduction ensures diversity, but there is sameness among individuals reproduced asexually. Most of us treasure uniqueness, especially among family and friends, and survival demands heterogeneity—not sameness. Thus cloning, although a useful procedure to the horticulturalist and experimental biologist, is not an appropriate method for reproduction. It seems that 4 billion plus souls are enough for this fragile sphere. Cloners seek a better life for those who are already here.

Efforts to clone humans clearly are not in the mainstream of biological research. Human cloning attempts will be costly and difficult. Certainly no scientific question will be answered. However, human cloning presents a challenge, and some people respond to technical challenge just as others, like the mountain climber, respond to physical challenge. Therefore, it should not be surprising that occasionally someone will attempt human cloning. I trust that that individual, should he or she ever be successful, will experience great personal gratification for having perfected a difficult procedure. But he or she will add little or nothing to the immediate welfare of humans.

A theologian makes a modest error in biology—a science writer creates a concern that extends all the way to the Congress of the United States. James Dewey Watson, Nobel laureate in molecular biology, said, "It might be expected that many biologists, particularly those whose work impinges upon (cloning), would seriously ponder its implication, and begin a dialogue which would educate the world's citizens." This book is my contribution to that dialogue.

2

"A Fantastical Experiment"

We may be on the threshold of mammalian cloning. And it seems that the successful cloning of mammals would mean that there is at least the technical possibility that humans may some day be cloned. Since the cloning of humans raises enormous ethical and moral issues, and since it fails to heal any known disease, then it seems appropriate to ask the question that was asked in a medical journal: "Should we have taken the first step?" (Genetic Engineering: Reprise. Editorial, *The Journal of the American Medical Association*, 220:1356-57, 1972). The first step in the present situation probably refers to the experimental studies that have made the cloning of a human being at least a theoretical possibility.

Designating any particular experiment as the *first* step is entirely arbitrary. Several late nineteenth-century experiments were instrumental in providing new insight into the cell biology of development—and they ultimately provoked the nuclear transplantation experiments. Thus, the first step(s) in animal cloning were taken years ago, and these studies are the substance of classical embryology—classical not because it is ancient, but because it led directly to the research programs of late twentieth-century cell and molecular biology. To ignore or bury these studies would be to obliterate the foundation of contemporary biomedical progress which holds the promise for new insight into such maladies as cancer and aging.

THE "FIRST STEP(S)"

Nineteenth-century biologists sought to determine if cells of an embryo develop independently or if they interact and affect the fate of each other during embryogenesis. Is the early embryo a mosaic of autonomous cells developing according to intrinsic control and producing a population of diverse and differentiated cell types that will comprise the adult, or is the early embryo a mass of cells that influence one another and that, by coordinated effort, produce an adult constituted of the various cell types? The early experiments relate not only to embryonic cells but also to cell components—nucleus and cytoplasm.

A fertilized egg, also known as a zygote, consists of a nucleus and cytoplasm (Figure 1). In fact, almost all cells—brain cells, liver cells, skin cells, as well as zygotes—are composed of nuclei and cytoplasms. Nuclei contain DNA (deoxyribonucleic acid), the hereditary or genetic material of the individual. The cytoplasms contain the biochemical and physiological support systems for the maintenance of the cell and the individual. One need not take a course in histology, the study of microscopic anatomy of tissues, to comprehend that the anatomy of a brain cell is different from that of a liver cell and the liver cell structure differs from that of a kidney cell, skin cell, etc. (Think of the differences in the texture and color of brain and liver in the supermarket.) The term "differentiation" refers to the developmental events that lead to these obvious differences, to the

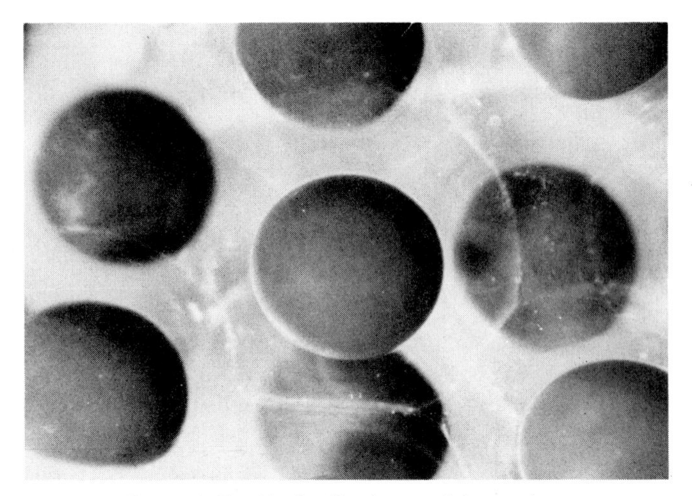

Figure 1. Freshly fertilized eggs of the northern leopard frog, *Rana pipiens*. Each egg is contained in a transparent jelly membrane

many cell changes that occur as development proceeds from the zygote to the mature organism. It is in the mature organism that component tissues manifest their unique structures and functions.

In human beings and other higher animals, the differentiated state is extraordinarily stable. If a person is fortunate enough to live to be 90, his or her liver remains liver, kidney remains kidney, skin is stabilized as skin. The extraordinary stability of the differentiated state, which we take for granted, was the subject of the investigations of pioneer biologists.

August Weismann (1834-1914), a German who taught zoology and comparative anatomy at Freiburg, published in 1892 a landmark theory relating to the nature of the differentiated state. Weismann's theory is no longer tenable, but he has a memorable position in the history of embryology because of the questions he raised. Weismann postulated that the fertilized egg contains all the genetic determinants to form a complete individual. Thus the nucleus of a zygote, which contains the genetic material DNA, should have all the genetic material to form an entire individual, since an entire individual results from the development of the zygote.

The zygote first divides into two cells known as blastomeres (Figure 2). Blastomere, derived from Greek, means "part of a bud." The fertilized egg is the unopened bud of a new individual. When that bud divides into two, then four, then eight, then more cells on its way to flower as a new adult, the early cells are, of course, parts of the bud.

Weismann supposed that the genetic determinants of the zygote are divided when the egg divides. Thus the right blastomere of a two-cell stage would contain the genetic determi-

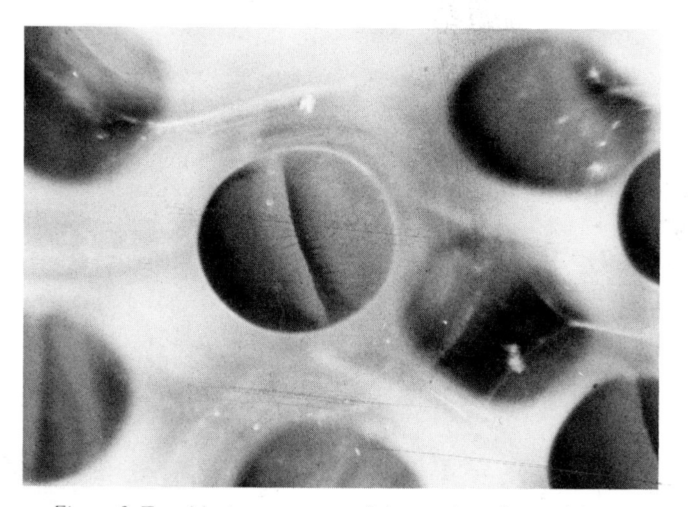

Figure 2. Two-blastomere stage of the northern leopard frog, *Rana pipiens*. An egg divides into two blastomeres about 3.5 hours after fertilization at 18°C

nants (genes or DNA) to form the right half of an embryo, and the left blastomere would contain the genetic determinants to produce the left. When the two-blastomere stage gave rise to the four-blastomere stage, the genetic determinants would be divided so that each of the four cells would contain one-fourth of the DNA originally present in the zygote nucleus. This process would continue sequentially until the liver was structured with cells containing genetic determinants (DNA) only for liver, the brain-cell nuclei with genetic determinants only for brain, and skin nuclei DNA that is specific for skin. Weismann's hypothesis certainly seemed reasonable because of the extraordinary stability of the differentiated state. The person who lived to be 90 had liver that remained liver because the genetic determinants contained in liver cells could specify for *nothing but liver*—or so Weismann believed.

Hans Spemann, who will be discussed later, was a pioneer German scientist noted for the experimental analysis of amphibian development. He recognized that Weismann's theory was useful because it prepared the ground for experimental embryology and suggested experiments.

An early experiment related to Weismann's theory was performed by German embryologist Wilhelm Roux (1850-1924), who founded the first scientific journal concerned with the experimental analysis of development in 1894—it is still published. Toward the end of the last century (to be more precise, during the spawning period of the European edible frog *Rana esculenta* in the spring of 1887), Roux obtained fertilized frog eggs, waited until they reached the two-blastomere stage (Figure 2), and then killed one of the blastomeres with a hot needle. If Weismann was correct, the surviving blastomere would give rise to only half an embryo. A half embryo was the result, and it seemed to those who looked at his illustrations (Figure 3) that Roux's experiment did, in fact, confirm Weismann's theory. I will say more about Roux's experiment later.

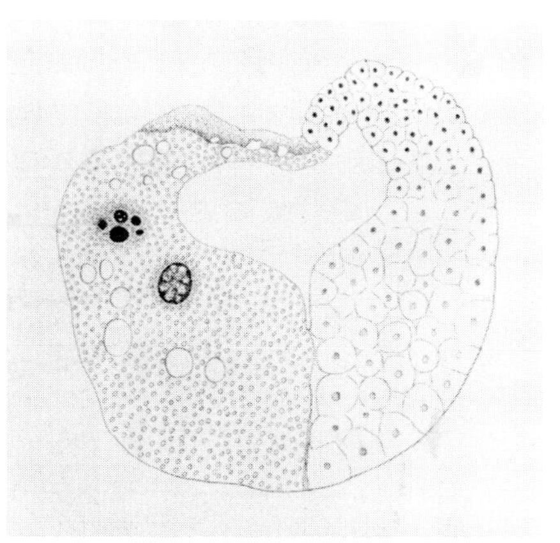

Figure 3. A half embryo develops following the death of one of the first two blastomeres. The embryo part on the right is cellular and the mass to the left is the remains of the dead blastomere. (From Roux, 1888)

About the same time that Roux was toiling with frog eggs, German embryologist Hans Adolph Eduard Driesch (1867-1941) was experimenting with sea urchin eggs. Sea urchins, small marine creatures related to starfish, are to this day exceptionally important in studies of fertilization, reproduction, and biochemical developmental biology. Driesch, in effect, performed the same experiment that Roux did, only in a different way. Sea urchin eggs are much smaller than the eggs of amphibians, so it would have been difficult for Driesch to destroy a blastomere by inserting a hot needle into it. Instead, he took sea urchin eggs at the two-blastomere stage, placed them in a flask, and shook the flask so vigorously that the two cells became detached from each other.

Driesch carefully recorded what happened to the separated blastomeres. They developed as whole embryos but they were dwarf. They were smaller than normal probably because the blastomeres were smaller than the initial undivided egg mass. But the most important event he observed was that they were whole embryos, not half embryos. It would seem that the philosophical implications of obtaining an entirety from a fragment were so astonishing that Driesch abandoned experimental embryology and became a professor of philosophy. In fact, manipulative embryology still has a philosophical impact in the late twentieth century.

Driesch's results obviously contrasted with Roux's. Driesch correctly observed that one important difference was that the blastomere treated with a hot needle in Roux's frog experiment was not *separated* from its compatriot. Driesch thought that the separation experiment ought to be done with fertilized eggs of higher animals. He attempted it with frogs and mourned, "I have tried in vain to isolate amphibian blastomeres, let those who are more skillful than I try their luck."

Among those who were successful was the late University of Minnesota Professor Jesse Francis McClendon, who reported in 1910 the development of isolated blastomeres of a frog egg. Gudrun Ruud in 1925 was equally successful with salamander eggs. McClendon, Ruud, and others reported that at the two-blastomere stage, under certain conditions, each of the blastomeres when completely separated from the other, would develop into a whole, intact embryo. Thus vertebrate embryos (vertebrates, animals with back bones, include frogs, toads, and salamanders, as well as humans) seemed to have the same capacity to develop from separated blastomeres as did sea urchin embryos. The conclusion we draw from these pioneer experiments is that, at least in the earliest part of development, genetic determinants are *not* divided among the blastomeres as Weismann postulated.

Let's return briefly to Roux's experiment involving the insertion of a hot needle into one blastomere at the two-cell stage of a frog embryo, with a half embryo resulting. Roux was a good reporter who failed to draw the right conclusions. The most accurate explanation of Roux's results seems to be that the mass of the dead blastomere, in intimate contact with the living cell progeny of the surviving blastomere, physically prevented movements of cells and otherwise inhibited the viable half from fully expressing its genetic potentialities. Blastomere *separation* is a better test of the developmental capabilities of a frog-embryo fragment.

Thus with sea urchins and frog embryos there seems to be no diminution of genetic potentialities when the egg divides into blastomeres at an early stage. However, an adult is composed of millions upon millions of cells. Is there any way to ascertain what the genetic capabilities are of nuclei obtained from older embryos or adults?

PRIMITIVE CLONING

Jacques Loeb (1859-1924) studied parthenogenesis, the development of unfertilized eggs. He used various salt solutions to stimulate development of virgin eggs. In some fertilized eggs subjected to this unnatural shock, the cell membrane tears and egg cytoplasm protrudes. In effect, this is a herniated egg—a nucleated mass (nucleus and cytoplasm) with a small appendage of non-nucleated cytoplasm. The nucleated cytoplasm divides when the zygote nucleus divides. The appended bleb, bereft of a nucleus, fails to divide for a time. Eventually, however, a nucleus may move across the cytoplasmic bridge into the formerly non-nucleated bleb. This movement of the nucleus, described by Loeb in 1894, is a nuclear-transplantation experiment, *a cloning experiment of nature* made possible by the diluted seawater. Sometimes the cytoplasmic protrusion becomes another embryo after nucleation, a twin of the first embryo. The important biological message of this spontaneous nuclear transplantation is that the genetic determinants of the zygote nucleus are not partitioned during cleavage divisions of the sea urchin embryo, which would have happened had Weismann been correct.

Hans Spemann (1869-1941), Professor at Freiburg and Nobel laureate referred to earlier, was aware of Loeb's experiments and wanted to extend the observations to vertebrate embryos. His method involved the constriction of a salamander zygote with a noose made of baby hair. (Baby hair was a valuable commodity in embryology laboratories at the turn of the century because very fine monofilament line was not available and baby hair was the finest and strongest thread at hand). The zygote became dumbbell-shaped as the baby-hair noose was gently tightened. The salamander zygote nucleus was in one portion of the dumbbell-shaped egg. The other portion was nucleusless, not unlike the herniated cytoplasmic protrusion of the sea urchin embryo described by Loeb. The nucleated portion of the egg divided into two, then four, then eight, then sixteen cells. At about this time, Spemann gently loosened the noose sufficiently so that one of the nuclei could traverse the cytoplasmic bridge to the unnucleated portion of the egg. When this occurred, that portion of the egg began to divide and the two egg fragments were then separated completely.

According to Weismann, the nucleus that invaded the egg fragment that was previously unnucleated would contain only a fraction of the genetic determinants of the zygote, and, therefore, the second embryo would develop imperfectly. Instead, development was normal but somewhat delayed—delayed because the secondary embryo got a later start (Figure 4). The significance of this experiment is that both fragments developed as whole, complete embryos. Weismann was wrong.

Spemann contemplated these results and wished that he could place the nucleus from a more differentiated cell in the cytoplasm of an egg deprived of its nucleus. He referred to this postulated manipulation as a "fantastical experiment." Spemann could envision the isolation of a nucleus; he could envision a fragment of egg cytoplasm devoid of a nucleus; at that time, however, he could see no way of introducing the isolated nucleus into the egg cytoplasm.

TOOLS FOR CLONING

Although Spemann could foresee no way of inserting a nucleus into egg cytoplasm in 1938, within only 14 years, his "fantastical experiment" was a fait accompli, performed by Robert Briggs and Thomas J. King in Philadelphia. The apparatus necessary for cloning, needing only minimal modification for use in nuclear transplantation, had long been present when Spemann's book was published in 1938. Thus Spemann's frustrations were unwarranted.

Problem solving is always simpler retrospectively than pro-

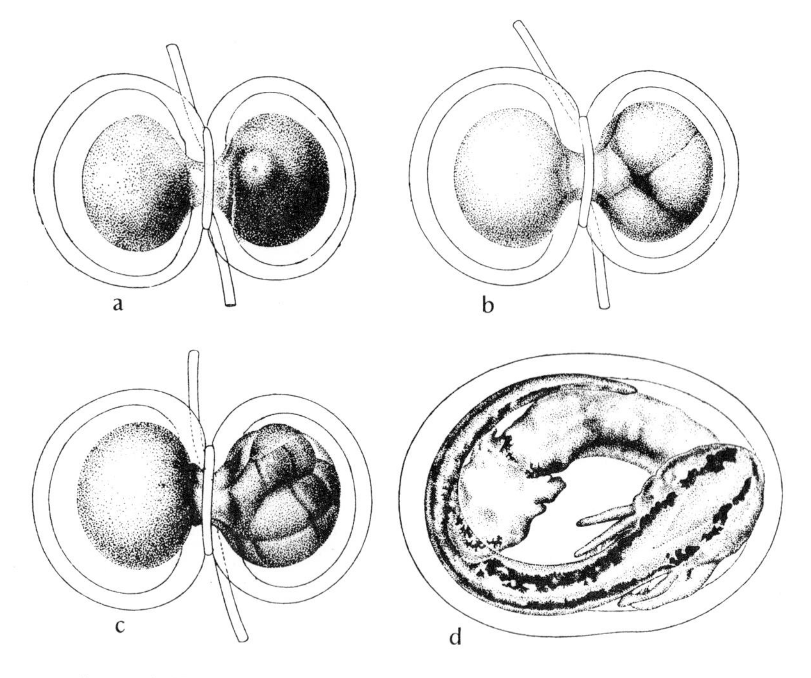

Figure 4. Twin embryos developed when a fertilized egg was constricted with a loop of baby hair in an experiment designed by Hans Spemann to test the developmental equivalence of early embryonic nuclei. (a) A fertilized egg has been constricted with a loop of baby hair. Only the egg cytoplasm on the right of the loop contains a nucleus. (b) Cleavage into four blastomeres occurs in the nucleated portion of the constricted egg. (c) A nucleus is permitted to move into the previously non-nucleated portion of the egg cytoplasm to the left at about the 16-cell stage. Following this, the previously non-nucleated cytoplasm begins to divide with the result that (d) two whole and normal embryos result. The embryo to the upper left is somewhat younger than the embryo to the lower right because this embryo started its development several hours later. (From H. Spemann, "Die Entwicklung seitlicher und dorso-ventraler Keimhälften bei verzögerter Kernversogung," *Zeitschrift für wissenschaftliche Zoologie*, 1928, 132:105-34, Figures 17, 19, 21, 23)

spectively. But how else in 1938 could a nucleus be placed in egg cytoplasm except by injection with a very fine diameter glass tube? (More recently, viruses have been used to assist in nuclear transplantation—but more of that in a later chapter.) The injection of material into cells requires a tube that is small enough to penetrate the cell without permanently damaging it, but large enough to accommodate whatever is being injected. The injection procedure is enhanced if there is an instrument to grasp and to move the fine glass tube with precision.

Glass tubes for microinjection and instruments to hold the tubes were available in 1938. L. Chabry, a nineteenth-century biologist who pioneered studies of separated blastomeres before Driesch, described capillary tubes suitable for experiments under the microscope in 1886. McClendon, whose isolated-blastomere experiments were referred to earlier, developed a micromanipulator that permitted microsurgery on eggs firmly held by the apparatus. McClendon described his apparatus and microsurgery in 1907 and 1908. In 1914 Marshall A. Barber, then at the Bureau of Science in Manila, described a microinjection apparatus. The Barber equipment was known to American cell biologist Robert Chambers. Chambers published several studies during the second decade of this century that, among other things, described the microdissection of living cells and the "sucking" of a nucleus into a capillary tube of fine bore. The nucleus was disposed of by "blowing it out." By 1931 a new machine was devised for the rapid production of microneedles and microcapillary tubes. The machine, developed by Delafield DuBois of Washington Square College, was adopted for cell studies by Chambers.

Chabry, McClendon, DuBois, and Barber were not alone. J. Comandon and Pierre de Fonbrune of the Pasteur Institute,

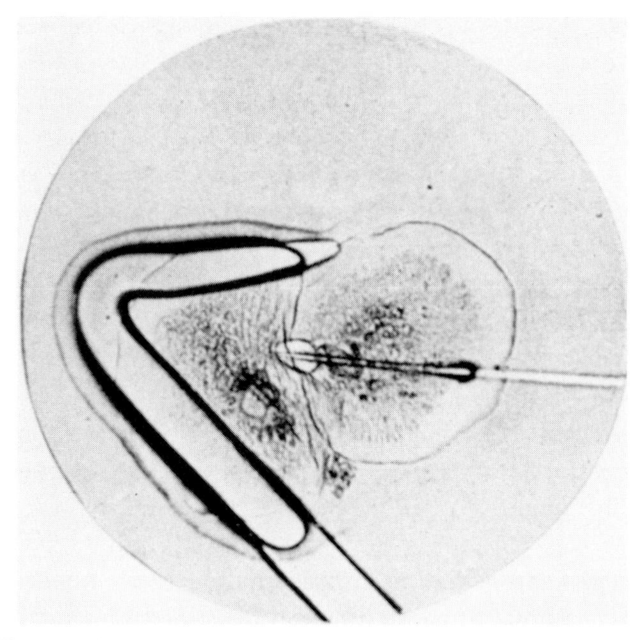

Figure 5. Nuclear transplantation in an amoeba. The recipient amoeba (left) is held in position with a glass hook. The nucleus of the donor amoeba (right) is pushed gently into the recipient cytoplasm with a glass needle. (From de Fonbrune, 1949)

Paris, described the manufacture of micropipettes, microneedles, and other microscopic glass tools during the 1930s. Not only were they engaged in tool making, but they were enucleating and transplanting nuclei in amoebae (Figure 5). Because of the relatively long history of experiments with cell microsurgery, it seems curious that Spemann, an internationally recognized Nobel laureate scientist, seemed unaware of how to get a nucleus into an enucleated egg.

Vegetative propagation, a form of plant cloning, is old. Animal biology always lags behind botany. However, by the late nineteenth century, the question that led to animal cloning was posed by Weismann and others, and preliminary answers were provided by blastomere separation. Better answers were forthcoming with observations of herniated eggs and constricted zygotes. It was but a brief pause before Briggs and King reported from Philadelphia their first successful nuclear transplantation experiments in 1952.

Should we have taken the first step? Does the question refer to man's manipulation of plants for over four millennia? Does it refer to separated blastomeres? Or the primitive nuclear transplantation experiment of a constricted zygote? Or the pre-World War II experiments with amoebae in Paris? These early steps comprise an important portion of the history of experimental embryology.

3

To Clone a Frog

The historical account of the experiments that led to nuclear transplantation explains why cloning was attempted. It does not reveal *how* cloning was accomplished. An account of the nuclear-transplantation procedure, "the how of cloning," provides insight into the purposes, opportunities, and limitations of the cloning technique.

Nuclear transplantation refers to the process of moving a nucleus from one cell to another. The transfer of a nucleus would have little biological meaning if it were not from one *kind* of cytoplasm to another *kind* of cytoplasm. Why was *egg* cytoplasm chosen as the recipient cytoplasm for inserted nuclei? Why were *blastula* nuclei studied first?

EGG CYTOPLASM

If one can enucleate an amoeba and place into that amoeba the nucleus from another amoeba, then there is little, if any, theoretical reason why an embryonic nucleus could not be inserted into the cytoplasm of a specialized cell. However, pioneer investigators chose to place an embryonic nucleus into enucleated egg cytoplasm. Their choice of egg cytoplasm was related to the fundamental question posed by these investigators. Do irrevocable changes occur in the genetic material of cells as they develop? Only if the transplanted nucleus is placed in egg cytoplasm can this question be answered.

A frog zygote nucleus in frog egg cytoplasm results in a developmental system that forms a frog. A human zygote nucleus in human egg cytoplasm results in a developmental system with the potentialities of forming a human. Egg cytoplasm has the support systems to permit programming of the zygote nucleus that results in an intact whole individual, be it frog or human. Consider, if you will, transplanting an embryonic frog nucleus into the cytoplasm of an enucleated brain cell. It requires little imagination to realize that brain-cell cytoplasm probably does not have an adequate supply of stored energy to allow for the development of a frog. Probably more crucial is the possibility that the cytoplasm emits signals to the nucleus that *preclude* expression of many genetic characteristics. This possibility seems particularly compelling as more is learned about cytoplasmic control of genetic activity.

Egg cytoplasm would be an inappropriate test site for the study of the capacity of a nucleus to guide development if the cytoplasm were preprogrammed for substantial development. Therefore, it is important to ask how much can egg cytoplasm develop without a nucleus?

Unfertilized eggs of many species develop rather well. As noted, Jacques Loeb obtained swimming larvae from unfertilized sea urchin eggs treated with seawater and magnesium chloride, proving that fertilization is not necessary for early development in this species. The chromosomes of the egg can sustain substantial development.

Will eggs deprived of sperm and bereft of their own chromosomes develop? J. F. McClendon provided some early observations on the behavior of eggs with no hereditary material. McClendon, in 1908, removed the maternal hereditary material from starfish eggs. The enucleated eggs were then stimulated with carbonated seawater. Development was initiated and a small cluster of non-nucleated cells was formed (Figure 6).

Ethyl Browne Harvey, an embryologist at Princeton University, continued and extended the studies of eggs that develop without genetic material. Not unlike McClendon, Harvey treated non-nucleated fragments of sea urchin eggs with chemical

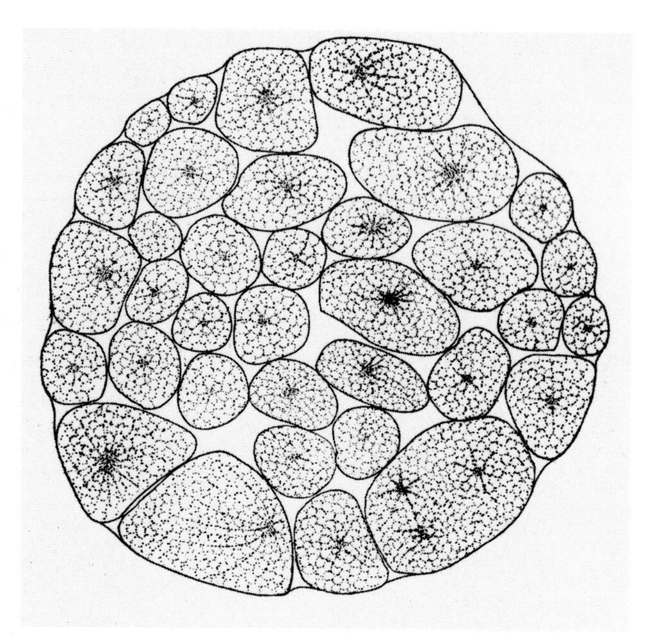

Figure 6. A starfish egg deprived of its nucleus is capable of forming a cluster of non-nucleated "cells." No differentiation ensues. (McClendon, 1908)

agents that stimulate development. A sea urchin egg, lacking *both* maternal and paternal hereditary material, when stimulated, will divide and form a ball of cells. No muscle, no nerve tissue, no skeleton forms. Cell division, but no cell specialization occurs in these curious non-nucleated sea urchin embryos. Sea urchins are a less complex form of life than frogs. Will frog eggs develop if they have no nucleus?

A haploid individual is one with chromosomes from only one parent. Haploid frogs develop regularly to the swimming tadpole stage but ordinarily no further. Recently, however, it was reported in Japan that a haploid tadpole had undergone metamorphosis to become a juvenile frog. A diploid individual contains chromosomes from both male and female parents. In the laboratory diploid development ordinarily follows when sperm are spread across freshly ovulated frog eggs. About 40 years ago, American embryologist and cell biologist Keith Porter noted that maternal chromosomes approached the surface of the frog egg shortly after sperm were placed on the egg. The maternal chromosomes were thus accessible to a microneedle. Porter removed the maternal chromosomes with the microneedle, and the surgically treated egg developed as a haploid. This kind of haploid development is known as androgenesis, meaning that only the paternal chromosomes participate. It is, then, relatively simple to produce a tadpole that develops with only one set of chromosomes. However, the question posed was will a frog egg develop with *no* nucleus, with no chromosomes whatsoever?

The answer came in an interesting study by Robert Briggs, Elizabeth Ufford Green, and Thomas J. King. These investigators treated sperm with toluidine blue, an anti-bacterial dye that is used to treat hemorrhage and that has an affinity for nuclei.It will stain the nucleus of a sperm, inactivating the hereditary material of the sperm without doing much damage to the sperm's motility or its capacity to penetrate an egg. What happens when an egg "fertilized" with a sperm that contributed no chromosomes is enucleated according to Porter's procedure? The egg, "fertilized" with sperm containing no functional *paternal* chromosomes, is then surgically treated for removal of the *maternal* chromosomes. The egg cleaves. It forms two blastomeres, four blastomeres, eight blastomeres—it forms a ball of cells, and the final stage is an imperfect blastula. Since tissue differentiation begins *after* the formation of a blastula, nucleusless frog eggs divide but they do not differentiate.

We now have our answer to the question of why egg cytoplasm was (and is) used. Egg cytoplasm deprived of its own chromosomes seems to be an ideal environment in which to test the differentiative capabilities of an embryonic or adult nucleus because it does not have the capacity to differentiate.

BLASTULA NUCLEI

Why did early nuclear transplanters clone nuclei derived from blastulae? Why didn't they transplant nuclei from older embryonic stages? There are two good reasons: one is that blastula cells are considered to be undifferentiated. Spemann, who did the primitive nuclear-transplantation experiments with constricted zygotes, exchanged patches of cells between young embryos. Even at the early gastrula stage (one stage later than the blastula), cells destined to form brain if left undisturbed form skin when transplanted to skin-forming areas, and cells destined to form skin when left in place form brain when surgically inserted in that portion of the embryo where brain is formed. Results are different when similar exchanges of tissue are made between older embryos. Late gastrula cells destined to form brain become brain even when transplanted to an anatomically inappropriate site. These observations, as well as many others, led embryologists to believe that blastula cells are not yet differentiated, have not yet developed specialized cell function. Thus these cells are appropriate subjects when one seeks to characterize the developmental potentialities of a kind of nucleus, when one attempts to see if the nuclear-transplantation system works.

If older, differentiated nuclei were chosen for cloning, and if development was imperfect, investigators wouldn't know if the imperfections of the experimental embryo resulted from the kind of nucleus that was transplanted or from the transplantation procedure.

The second reason for studying blastula nuclei is that they have already undergone twelve or thirteen cell generations. This may not sound impressive, but these generations produce a blastula composed of several thousand cells (calculate 2^{12} or 2^{13}). The ability of investigators to characterize the genetic determinants of a blastula nucleus was an enormous stride forward beyond Spemann's 16- or 32-cell-stage constricted egg experiment. Substantial new information became available when the nuclear transplantation of blastula nuclei was successful.

Validity of Blastula Nuclear Transfers

Robert Briggs and Thomas J. King reported to the scientific community in 1952 that normal tadpoles developed from their nuclear-transplantation experiments. How is the validity of such an experiment determined?

To Clone a Frog

A valid cloning experiment produces a frog asexually. No gamete nucleus, either sperm or egg, participates in the development of a frog that is truly a clone. There must be convincing evidence that sex-cell nuclei did not participate in the formation of nuclear-transplant frogs. The experiments described in the following discussion provide compelling evidence that cloned frogs are indeed cloned.

Nuclear transplantation requires that the recipient egg be enucleated. Aristotle knew that bee drones were produced without copulation. Loeb produced fatherless sea urchins with seawater and magnesium chloride. Tadpoles of frogs result after parthenogenetic stimulation of unfertilized eggs. There are at least two dozen species of reptiles that reproduce without males. Hence, extraordinary care was taken to ensure that maternal chromosomes were eliminated from the recipient, unfertilized egg, to exclude the possibility of parthenogenesis. Scientists wanted to determine what the developmental potential of the *inserted nucleus* was. They already knew the capacity of maternal chromosomes functioning alone.

As noted earlier, the production of androgenetic haploids involves removing the *maternal* chromosomes. The skill of haploid production witnesses to the ability of experimenters to recognize and surgically remove maternal chromosomes. Briggs and King reported that all the surviving tadpoles, in which they attempted to remove the maternal chromosomes from fertilized eggs, were haploid. Thus they indeed demonstrated their ability to obliterate the maternal hereditary apparatus. In their cloning experiments, in contrast to the haploid enucleation manipulations, the recipient eggs *were not fertilized*. Hence, the eggs could be considered nucleusless before the nuclear transplantation procedure.

If development that occurs after nuclear insertion is attributable to the transplanted nucleus (and not to an inadvertently retained set of maternal chromosomes), then that development should be constrained to follow the capacity of the donor nucleus and, as such, is nuclear specific. Hybrids have proved useful in establishing the specificity of nuclear development.

Hybridization has been studied extensively in amphibia. Since eggs and sperm of frogs and toads are readily available to experimenters, it is relatively simple to produce a hybrid by combining sperm from one species with eggs from another. Some combinations result in viable hybrids, others in lethal hybrids. Certain lethal hybrids are interesting to biologists because they develop to a particular embryonic stage and then development is arrested. The experiment that follows utilized the arrested development of a hybrid.

I have said that if development of the recipient egg is a function of the nucleus inserted into the egg, then development is limited by the kind of nucleus that is inserted. The hybrid formed with the sperm of a bull frog (*Rana catesbieana*) and the egg of a leopard frog (*Rana pipiens*) arrests at the end of blastulation or the onset of gastrulation. As noted earlier, many profound cellular and biochemical changes occur in the gastrula stage. What happens when the nucleus of the hybrid (bullfrog sperm crossed with leopard frog eggs) blastula is injected into an enucleated egg of a leopard frog? The cloned embryo invariably arrests at the onset of gastrulation. Control nuclear transfers, that is blastula nuclei from non-hybrid donors inserted into the appropriate egg cytoplasm, develop normally. Thus the nuclear-specific pattern of development is substantial genetic evidence that growth of the operated eggs was truly attributable to the inserted nucleus.

To Clone a Frog

There is a way of permitting normal development of the cloned embryo *and* having direct genetic assurance that development of that embryo is due to the hereditary material inserted. In Minnesota and some contiguous states, there are pigment mutants of the common leopard frog and, of course, the ordinary spotted variety (Figure 7). Two of these mutants are known as Kandiyohi (Figure 8) and Burnsi (Figure 9). Nuclei from the mutants have been used as genetic markers in cloning experiments.

The hereditary material of ordinary leopard frogs dictates that the developing frog will have leopardlike spots. Another

Figure 7. The northern leopard frog, *Rana pipiens*, with large and distinct spots on its body and legs. This is the species that was used in the first successful cloning experiments. (Photograph by Gordon A. F. Dunn)

Figure 8. The mottled Kandiyohi mutant of the northern leopard frog, *Rana pipiens*, that occurs principally in Minnesota. (Photograph by Gordon A. F. Dunn)

Figure 9. Northern leopard frogs with reduced or no spotting are known as Burnsi mutants and occur principally in Minnesota. (Photograph by Gordon A. F. Dunn)

way of stating this fact is to say that expression of the spotting patterns of frogs depends upon nuclear function. This fact is no less true of mutant variants than it is of ordinary spotted frogs. Hence it is possible to use the activity of the mutant gene as a "marker" for a transplanted nucleus.

If nuclei from Kandiyohi or Burnsi mutant blastulae are inserted into enucleated eggs of ordinary frogs, the development of the resulting nuclear-transplant frog is determined by the nuclear donor, not by the genome of the female that provided the recipient egg. Thus the nuclear-transplant frogs will have the pigment pattern of either the mottled Kandiyohi nuclear donor (Figure 10) or the near-spotless Burnsi nuclear donor, not the pigment pattern of an ordinary leopard frog. Experiments with Minnesota mutant frogs offer compelling genetic proof that the inserted nucleus provides the genetic instructions for the growth of the frog.

Other pioneer nuclear transplanters used different genetic tags or markers associated with the donor nucleus to provide persuasive evidence of the validity of cloning. For example, nuclei of most cells of the South African clawed toad, *Xenopus laevis*, contain two minute structures known as nucleoli. There is a mutant that has cells which usually have only one nucleolus. When the nucleus of a one-nucleolus mutant is inserted into an enucleated egg of a two-nucleoli female *Xenopus* and the resulting tadpole has only one nucleolus per cell, then we are justified in believing that it was the inserted nucleus, not an inadvertently retained maternal egg nucleus, that determined the development of the tadpole. Extra sets of chromosomes, known as polyploidy, have also been used as nuclear markers in frogs and salamanders. In all experiments with nuclear markers, the conclusion that follows careful observation of experimental em-

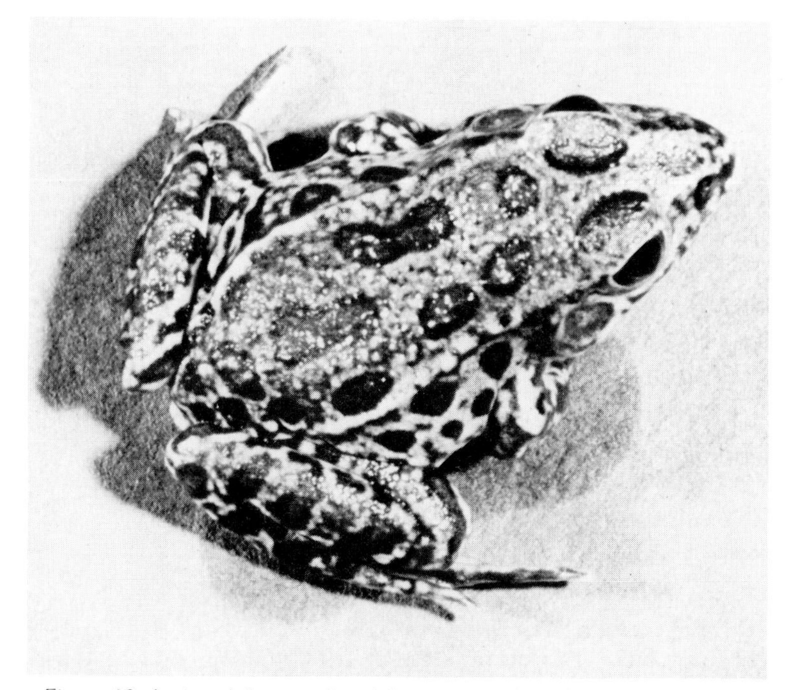

Figure 10. A cloned frog produced from an enucleated egg of a Vermont spotted leopard frog and an inserted nucleus of a Minnesota Kandiyohi mutant embryo donor. The expression of the mutant characteristic (Kandiyohi) is genetic evidence of the authenticity of the cloning procedure

bryos and frogs has been that it is the donor nucleus that guides or programs development. Because it is a body-cell nucleus, not a gamete nucleus, the cloned embryo is asexually produced.

The enucleation procedure is valid, and there is commanding evidence of several kinds that provides assurance that the inserted nucleus guides development. So, if the cloning procedure works for the American leopard frog, can it work with other species? In the years following the prototypic Philadelphia cloning experiments, the technique was extended to many different amphibian species. Already mentioned was the South African clawed toad, *Xenopus laevis*. To this may be added various species of frogs and toads from America, Europe, and Asia, as well as salamanders. It may be interesting to note that nuclear transfer has also been accomplished in the fruit fly, *Drosophila*, and cloning studies in fish are under way in the People's Republic of China. Thus far, frogs, toads, and salamanders remain the most studied cloned animals.

Nuclear transplantation results are believable because a variety of genetic manipulations indicate that the development observed in cloned animals is authentically attributable to the inserted nucleus. How, then, is cloning done?

THE CLONING PROCEDURE

How difficult is the cloning procedure in frogs, toads, and salamanders? It is very difficult, but far simpler than mouse and rabbit cloning. This fact has certain implications for the plausibility of claimed human cloning—which will be discussed later. Cloning amphibia was successful a quarter of a century ago but mammalian cloning is not yet successful. The early success with frogs may be attributed to a century of experimentation in the reproductive biology of amphibians. Before the first successful frog-cloning experiments were performed, a number of procedures had been developed that would become useful to the craft of nuclear transplantation. These included the ability to obtain eggs and sperm from frogs, in vitro fertilization, removal of maternal chromosomes from eggs, and dissociation of embryos into individual cells.

To Clone a Frog

Obtaining Frog Eggs for Cloning Experiments

The northern leopard frog, *Rana pipiens*, spawns in the spring, and no eggs can be ovulated immediately after spawning. The frog spends the summer months foraging in the fields and growing eggs. By the time fall arrives, the eggs have grown to their maximum size and the frogs are ready for hibernation under the ice of lakes and streams. Athough egg release will not occur spontaneously until the next spring, ovulation can be induced from September to or past the time of natural ovulation. Eggs leave the ovary, move to the reproductive tubes, and become available to the embryologist when the female frog is injected with pituitary glands or a combination of pituitary glands and the hormone progesterone. The eggs can be extruded from the female after this treatment by gently squeezing the abdomen.

Obtaining Frog Sperm

Sperm can be obtained by cutting frog testes into fine pieces in a diluted salt solution. The testes (testes is plural for testis, the male sex-cell gland; testicle is a diminutive and not a particularly appropriate word for a sexually mature beast) are dissected from the male, which usually requires sacrifice of the frog donor. Sperm may be obtained without sacrifice by hormonal release. This procedure is used in my laboratory to spare the increasingly scarce frogs found in nature. Chorionic gonadotropin (a commercially available hormone present in pregnant humans) is injected into a mature male frog. Within one hour, motile sperm are released from the testes of the frog and are found in the urine. The sperm are capable of fertilizing frog eggs, and the experiment, in addition to being useful to the cloner, shows that reproductive glands of evolutionarily primitive cold-blooded frogs respond to hormones of evolutionarily advanced warm-blooded human females. This suggests the continuity of life and the similarity of biologic processes among the vertebrates since ancient times.

Fertilization and Husbandry of Frog Embryos

Eggs and sperm can be combined in a glass dish at a carefully predetermined time. By caring for the fertilized eggs at a particular temperature (usually 18°C) and a particular time, donor embryos of predetermined stage can be obtained.

There is no need to provide an elaborate culture medium for developing frog embryos, as one must do with the much smaller mammalian embryos. Simple glass dishes and water are adequate because frog eggs are large and contain stored food enough for about 12 days. Frog tadpoles are vegetarians when they begin feeding and survive nicely on cooked lettuce or spinach until they become carnivores at metamorphosis about 90 to 100 days after fertilization.

Preparing Eggs to Receive a Transplanted Nucleus

Freshly ovulated eggs, contrary to what many textbooks state, have the same amount of DNA as an ordinary body cell. That amount of DNA is twice the amount contained in a sperm; hence, it is called diploid. A sperm has the haploid amount of DNA. Diploidy in freshly ovulated eggs is a boon to the experimenter. If diploid eggs combined with diploid sperm, the amount of DNA of the resulting individuals would become enormous in only a few generations. This, of course, does not occur. What happens is that the final maturation of the frog egg, to become haploid as the sperm already is, occurs *after* it is released from the ovary and at that time it is activated by the penetration of the sperm.

Embryologists can *mimic* the activation of sperm penetration by pricking the surface of the freshly ovulated egg with an extremely fine glass needle. The maternal chromosomes respond to the needle prick by approaching the surface of the egg in what would ordinarily be an attempt to extrude half the chromosomes, thus half the DNA, in a minute cell known as a polar body. It is at this point that the maternal chromosomes and the chromosomal DNA of the unfertilized egg are accessible to either surgical removal or removal by the flash of a laser apparatus mounted on a microscope (Figure 11). Removal of maternal chromosomes and chromosomal DNA, whether by surgery or laser, results in an egg devoid of any genetic material in the form of chromosomes. The enucleated egg now needs only to be removed from its jelly envelope (by cutting with very fine scissors and forceps) to be a suitable recipient for a transplanted nucleus.

Preparing the Donor Nucleus for Transplantation

The eggs that were fertilized are permitted to attain an appropriate embryonic stage to serve as nuclear donors. This need not take long. An embryo composed of several thousand cells develops in less than 24 hours at 18°C.

When a donor embryo is selected, the jelly is removed and the appropriate region of the embryo is dissected and placed in

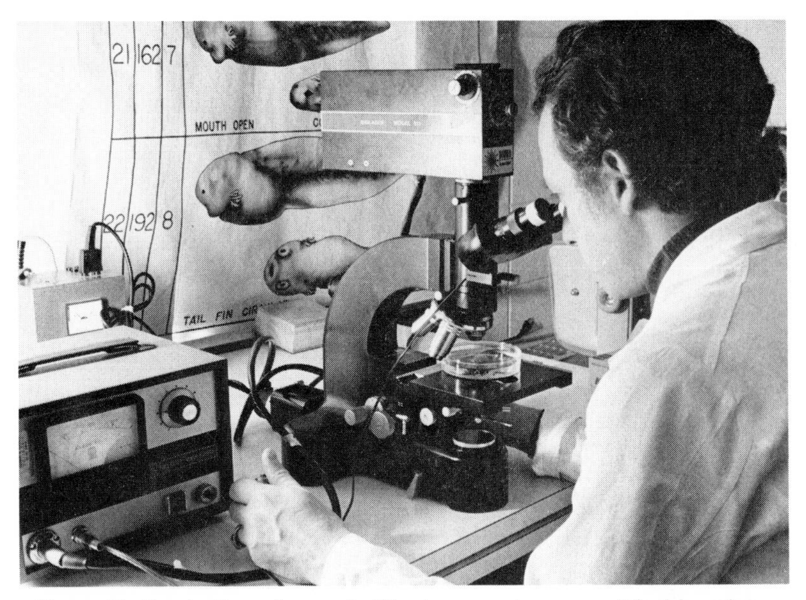

Figure 11. Enucleation of an unfertilized egg can be accomplished by microsurgery or by irradiation with a ruby laser apparatus mounted on a microscope

a solution that causes the dissociation or separation of individual cells. A physiological salt solution containing neither calcium nor magnesium salts is used. Young embryos in a calcium- and magnesium-free solution rapidly dissociate to individual free cells. With older embryos, it may be necessary to add special substances (such as agents that chemically bind free calcium in the solution or protein-digesting enzymes that break down the complex material that binds cells together into a tissue fabric) to the dissociation solution to obtain free cells.

It is relatively simple to bring together the separated donor cells that provide nuclei and the enucleated host eggs. The mechanics of the conjugal surgery involve micropipettes, microinjection apparatus, and micromanipulation equipment.

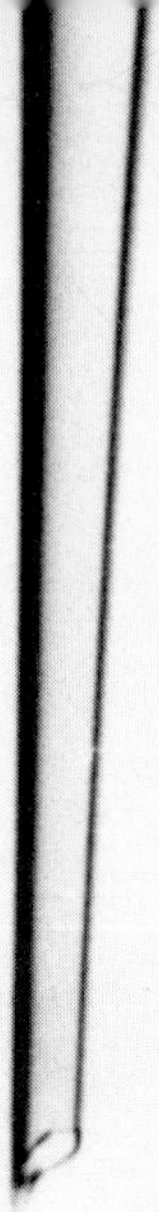

Figure 12. Glass micropipette with a beveled and sharpened tip. (From McKinnell, 1978)

A pipette is a small glass tube used to transfer liquids. A micropipette (Figure 12) differs from an ordinary pipette only in size (it is tinier) and acuteness of tip (it is sharper). A micropipette is selected with an orifice that is somewhat smaller than the diameter of the donor cell. The micropipette is positioned adjacent to the selected dissociated cell with the microinjection apparatus.

The microinjection apparatus (Figure 13) is simply a machine, a micromanipulator, that holds a tool (in this case a micropipette) very steady, and permits small and precise movements of that tool. The donor cell is drawn into the micropipette with

Figure 13. Nuclear insertion is accomplished with the aid of a microinjection apparatus and microscope. The scientist has her hand on a heavy-duty syringe which is connected to a micropipette held in position with a micromanipulator. The electrically focused microscope is foot-operated, which permits the scientist to use both hands in the cloning procedure

the microinjection apparatus. The procedure is like squeezing the rubber bulb of a medicine dropper in order to draw fluid into the dropper. When the cell enters the orifice of the micropipette, the cell membrane is ruptured (recall that a cell too large for the micropipette orifice was chosen), and there is some dispersal of the donor-cell cytoplasm. The cell membrane is very thin but very important. If it is left intact on the inserted donor cell, the enucleated ovum with its donor cell cannot develop. However, a donor cell with its nucleus liberated by virtue of a broken cell membrane can interact with the egg cytoplasm to form a viable developmental system—sometimes resulting in the formation of a frog.

The Skill Required for Nuclear Insertion

Cells are microscopic. Nuclei are microscopic. The maternal chromosomes are microscopic. The movements of the micropipette are microscopic. Thus the procedures just described must be performed under the relatively high magnification of a light microscope. Although cloning is conceptually simple, the operation is exquisitely complex; and much skill, coupled with precision, is required of the operator.

Cloners not only have to be expert at reproductive biology and cell surgery, they must also be proficient enough to make the implements of the craft. Probably the most difficult implement to make is a beveled and sharpened micropipette with an orifice appropriate for an embryonic cell. Consider the skill required to make an ordinary medicine dropper. Add to that skill the ability to change the orifice of the dropper to a sharpened bevel. Now consider the skill required to accomplish all of this but also to make the beveled and sharpened dropper so tiny that, when properly used, it can break a single cell while not damaging the nucleus contained within. That is why micropipette-making is exacting.

It is extraordinarily difficult to penetrate the surface of an enucleated egg with an improperly sharpened and beveled micropipette. The otherwise delicate egg has a membrane that is tough and very stretchable. A dull micropipette depresses the surface of the egg with pressure by the micromanipulator. The depression becomes so great that the improperly sharpened micropipette sometimes goes all the way through the egg and protrudes from the other side without having penetrated the egg membrane. Or the pressure exerted from a dull micropipette may rupture the egg membrane, and egg cytoplasm spews all over the operating dish. These are certainly unacceptable results and ordinarily do not occur with a finely sharpened micropipette.

After the nucleus is transplanted, there is little to do other than record growth and development. Frog husbandry, like the care of mice or any other laboratory animals, is expensive. It takes more than a year for a frog embryo to attain sexual maturity. However, despite the time and expense involved in rearing, a few cloned frogs have grown to sexual maturity in the laboratory (Figure 14).

SIGNIFICANCE OF TRANSPLANTS

What do the frogs formed from cloned blastula nuclei show? The frogs witness to the capacity of skilled embryologists to take apart an early embryo to the level of whole but separated cells, to break one of those cells in order to free its nucleus, to

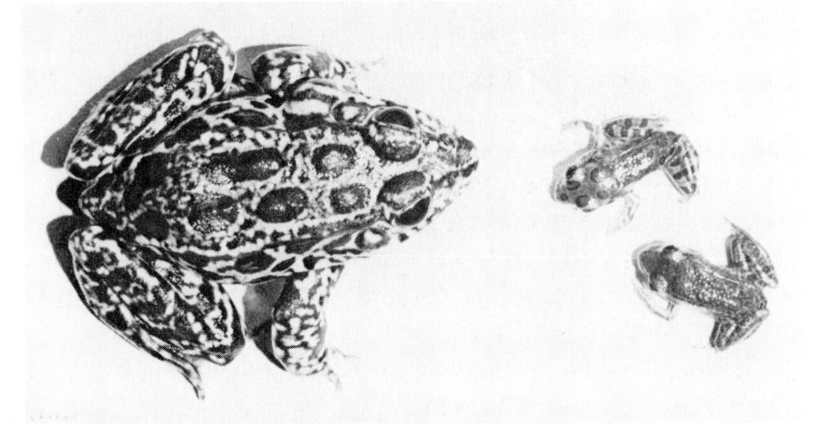

Figure 14. Sexually mature frog produced by cloning. The small frogs are two of many offspring fathered by the clone. (From McKinnell, 1962)

place that liberated nucleus into a previously prepared enucleated egg, and to effect the transplantation with such dexterity that the result is a viable embryo with the capacity to develop into an adult. More important than the skill of the embryologist is the biological significance of the cloned frog. The frog attests to the wholeness of the genetic material of the blastula nucleus from which it was contrived. This demonstrated intactness of blastula DNA is in harmony with the early experiments of Driesch, McClendon, Loeb, and Spemann on younger developmental stages.

From time to time, biologists, like most thoughtful people, ponder the nature of life. Is a single cell, derived from a vertebrate embryo alive? Of course. It will thrive in culture and give rise to many progeny. Is a nucleus, surrounded by cytoplasm, but not protected by an intact cell membrane, alive? Hardly. It will not divide. It will disintegrate in just a few hours. Is an unfertilized egg alive? Of course. Many can be stimulated to divide and give rise to individuals by parthenogenesis. Is an *enucleated* egg alive? Hardly. It will eventually disintegrate. Like the liberated nucleus, it has no independent life. However, let a skilled microsurgeon combine the not-alive liberated nucleus with the not-alive enucleated egg—and often a frog is formed.

Does the skilled cloner create life? No. The conditions that permit life are *restored* by cell surgery—and I suppose that is one reason why cloning is fascinating to experimenters. They seek answers to significant questions, and each experiment is a test of their ability to restore conditions that permit a spark of life to flame as a living substance capable of developing into a mature organism.

4

Cancer, Aging, and Other Challenges

The success of nuclear transplantation with blastula cells has encouraged researchers to do similar experiments with older, more specialized nuclei. Adults are composed of many kinds of specialized (differentiated) cells. If, as some believe, certain kinds of cancer result from the improper functioning of the differentiation process, then perhaps some cancer can be viewed as specialized cells that are harmful to humans. Humans age and the changes that occur in the cells of older people may be viewed as another kind of non-beneficial specialization. It is believed that the cloning procedure has the potential of revealing to what extent cell specialization is attributable to alterations in the genetic material. Therefore, a variety of specialized cells, including cancer cells, have been studied by cloning. A cloning study relating to aging cells has begun in my laboratory at the University of Minnesota.

There are four major categories of cloning research: differentiation—much has been done in this area of research that stimulated the early cloning experiments; cancer—a much studied problem; immunobiology—only tentative, but promising probes have been made thus far; and aging—exciting work which is in its early stages.

Cancer, Aging, and Other Challenges

Before discussing cloning experiments, let me point out that studies in genetics and cell biology can be conducted on several levels of complexity. An ordinary microscope with glass lenses and visible light has a maximum useful magnification of about 1000. It is possible, for example, using fixed and stained slides, to determine with an ordinary microscope whether nuclear DNA is fairly evenly distributed or is clumped. For fine anatomical detail, an electron microscope is needed. For example, the question of whether viruses are present in nuclei prepared for examination must be answered by the electron microscope, not the light microscope, because viruses are too tiny to be visualized with an ordinary microscope. A molecular biologist with centrifuge, electrophoresis apparatus, and other paraphernalia may wish to describe patterns of DNA replication in different kinds of cells and to gain insight into the nature of cell function by analysis of the replication patterns. I mentioned this not exhaustive inventory of cell-probe procedures to make a point. The fate of a cell and its nucleus is fixed when one prepares it for examination by conventional microscope or molecular procedures. Elegant as is the image obtained with the electron microscope, it can suggest only what the nucleus *was*. The molecular biologist analyzes an extract from cells and trusts that the findings have some relationship to the previously living cells.

The cloner, in contrast to some other cell biologists, asks what a nuclear transplant egg will become, not what it has been. The system is alive and the answers are provided by the living embryo that results from cell surgery. This is not to say that *all* cell biologists work with lifeless material. Tissue-culture studies of cell populations suspended in flasks of nutrient material, or growing as monolayers in flat dishes containing appropriate liquid cuisine, certainly relate to life. Details of the anatomy of these living cells in culture can be observed with a special microscope, the phase-contrast instrument, that does not involve killing cells in order to observe them. However, cells in culture do not form intact and functional embryos and adults. This is perhaps another reason why nuclear transplantation is fascinating to me and others. We do not surmise that a nucleus has certain capabilities, we do not conjecture that a nucleus may interact harmoniously with certain cytoplasm. We know this as we observe living, swimming, vertebrate embryos, some of which develop to the adult state.

CLONING STUDIES OF DIFFERENTIATION

More nuclear transplantations have been done with two amphibian species than with all other species combined—the North American leopard frog, *Rana pipiens*, and the South African clawed frog, *Xenopus laevis*. These frogs are as different from each other as frogs can be. Life-style: *Rana* is terrestrial in the summer, *Xenopus* is wholly aquatic; anatomy: the *Xenopus* tadpole is primitive, with soft mouth parts and two apertures for the exit of water from the gills; the *Rana* tadpole has hard mouth parts adapted for scraping, with but a single opening for water that has passed over its gills; mating: the male *Xenopus* clutches the female at the posterior portion of her body; the amorous male *Rana* embraces his female at the level of her front limbs. Ardent males are believed to reveal something of their evolutionary history in the mode of their hug, the *Xenopus* posterior clasp being primitive, *Rana*'s embrace, advanced. There are, of course, many other anatomical, physiological, and behavioral differences of greater interest to the herpetologist than to the general reader. The differences between primitive *Xenopus* and developed *Rana* are an important factor in nuclear transplantation because if the results of the cloning were not similar, this might be attributed to their substantial biological differences. However, if the results are similar, the concurrence at least suggests that the data are real and there is a possibility that similar results can be obtained with other species.

Results derived from cloning increasingly older embryonic nuclei are essentially the same in *Rana* and *Xenopus*. The success of nuclear transplantation decreases as the age of the donor nucleus increases (Figure 15). The yield of normally developing nuclear-transplant embryos drops precipitously as the *age of the donor nucleus increases beyond day one*. Unfortunate indeed is the fact that science-fiction writers and others who have expressed apprehensiveness about the results of cloning are unaware of this.

Investigators from several laboratories have attempted to transplant nuclei from adult cells. *No normal frogs have resulted from these studies.* Some tadpoles have been produced, all of which have died. But the tadpoles are exceptionally exciting to many cell biologists. The production of tadpoles in nuclear transfer from the nuclei of diverse adult cell types such as white blood cells, skin, and kidney cancer cells reveals that nuclear activities can be reprogrammed, at least to a limited extent. Total reprogramming, for the moment at least, is limited to nuclei from very young embryos.

Why are older embryonic nuclei and adult nuclei less capable than younger embryonic nuclei of promoting normal development in enucleated egg cytoplasm? This question relates to the primary problem of developmental biology—the differentiated state in all higher organisms. Modern cell biology maintains that

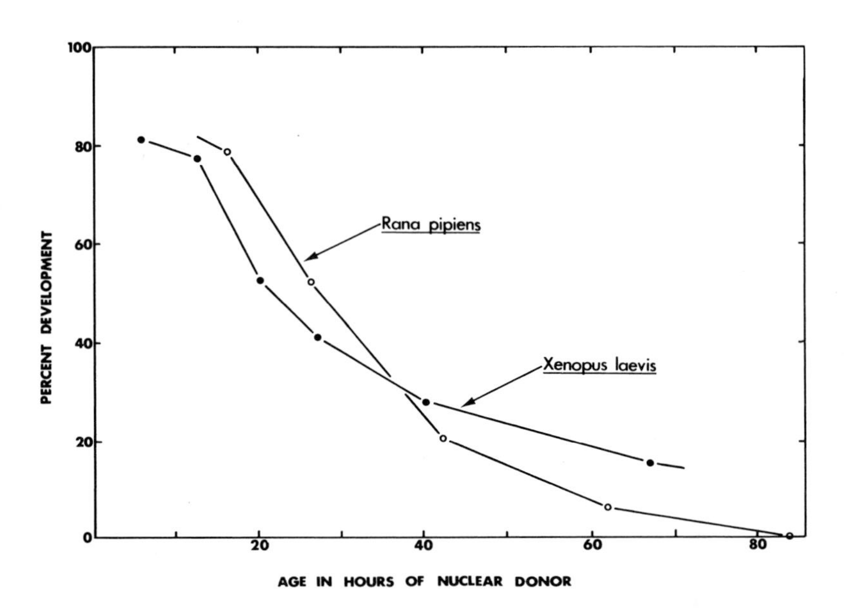

Figure 15. The capacity of a nucleus to participate in normal development, as tested with the cloning procedure, decreases rapidly as the age of the donor embryo increases. The majority of blastulae produced by transplanting nuclear donors less than one day old develop normally. Few blastulae produced from nuclear donors two days of age or older develop normally. (From R. G. McKinnell, "Nuclear transfer in *Xenopus* and *Rana* compared." In R. Harris, P. Allin, and D. Viza, eds., *Cell Differentiation* © 1972 Munksgaard International Publishers, Inc., Copenhagen, Denmark)

brain cells are different from liver cells, not because they have different genetic material, but because their common genetic material is organized differently or utilized differently. The reduced yield of normal frogs from more differentiated nuclear donors reflects the highly stabilized condition of the differential organization or use of genetic material.

Many scientists have speculated about why adult nuclei seem less capable than younger nuclei of promoting normal development when transplanted to egg cytoplasm. One possible explanation is that DNA, the basic genetic material, may be altered or rearranged during the course of development. The altered or rearranged genes *may* affect the capacity of a nucleus to be cloned. I do not know if changes in DNA structure are related to the failure to clone adult nuclei perfectly. However, if DNA changes in *many* differentiating cell types (and there is certainly evidence that it changes in *some*), then the cloning of adults (human and nonhuman) may be impossible for *genetic* reasons.

Microbiologists who study the controls of genetic activity in bacteria know far more about regulation than do scientists who work on similar problems in higher organisms. If more were known about gene regulation in higher organisms, it is probable that nuclei from maturing embryos and adults could be successfully cloned. Perhaps what keeps many nuclear transplanters at their laboratory benches is the belief that if they are ever able to treat an adult nucleus in such a way that it will program for a normal individual when cloned, this mode of treatment will reveal much about the differentiated state. And that is a primary concern of developmental biology.

CLONING AND CANCER BIOLOGY

Most readers know that many older people die of cancer. But many may not know that if a young person dies of a disease, the chances are greatest that the disease will be cancer. After several decades of intensive research, cancer remains a major medical problem in America.

Cancer related to environmental insult is receiving much attention. Monitoring air, water, and food for possible cancer-causing chemicals is becoming increasingly important with the recognition that noxious substances in our environment may cause cancer.

Although it is true that much cancer is related to environmental insult, it is also true that the origin of some (much?) cancer is not related to industrial chemicals, urban pollution, and cigarette smoking. Studies of fossil remains of pre-Columbian Peruvian and North American Indians show that a particular kind of bone cancer, multiple myeloma, was common. Certainly these unfortunate individuals were not exposed to automobile exhausts or chemical plant effluents. Such observations of cancer in prehistoric peoples suggest that even if the environment is cleaned of noxious man-made substances that are known or suspected to cause cancer, it seems probable that cancer will still afflict humankind.

Cancer has been with humans for years and is not likely to go away in the near future. How then are those who become afflicted with the disease to be treated? There are three principal treatments: surgery, radiation, and chemotherapy. Chemotherapy is a particularly significant kind of cancer treatment because of the dismal fact that about two-thirds of cancer patients have metastatic cancer at the time of diagnosis. Metastatic cancer is cancer that has spread via the blood or lymph to many parts of the body, many secondary tumors forming as colonies from the primary cancer. Surgery cannot eliminate these myriad malignant growths, nor can radiation—no one survives total body irradiation. Hence the need for chemotherapy.

Unfortunately, chemicals produced to kill cancer cells are exceptionally toxic. They are toxic because it is difficult to design chemicals that will distinguish between a normal cell that is dividing and a malignant cell that is dividing. If more were known about cancer-cell genetic material and about regulating its differentiated state, there would be the rational hope that a cancer chemotherapeutic agent could be devised that would regulate differentiation of these malignant cells.

The urgent need for a nontoxic agent to regulate gene expression in malignant cells is one reason why cloning experiments designed to reveal the nature of the genetic material of the cancer cell and the capacity of that malignant genetic material to give rise to normal cell progeny are being performed. Cloning may be the most direct method of acquiring information about these two important aspects of cancer-cell biology.

The experimental design of a cancer-cloning study is simple. Nuclei from a highly malignant tumor are inserted in a previously enucleated egg. If the egg divides, one expects the result to be either a ball of cancer cells or a tadpole of some kind. A ball of cancer cells would suggest that the cancer being studied is not easily altered from its malignantly differentiated state. A tadpole would argue that the genetic material of the cancer under scrutiny can be redirected to a less malignant state.

If a tadpole is produced by cancer nuclear transfer, indicating a reversal from malignancy, further experiments could be suggested that might ultimately lead to control of malignant gene expression. It seems relatively obvious to me that treating cancer by modifying gene expression promises to be less traumatic than treating cancer with cell-killing agents.

The leopard frog of North America, *Rana pipiens*, that served so well in the prototypic nuclear-transplantation experiments, may be afflicted with a relatively common kidney cancer (Figure 16). Balduin Lucké, Professor of Pathology at the University of Pennsylvania, who originally described the frog cancer, believed that the kidney cancer cells had a normal number of

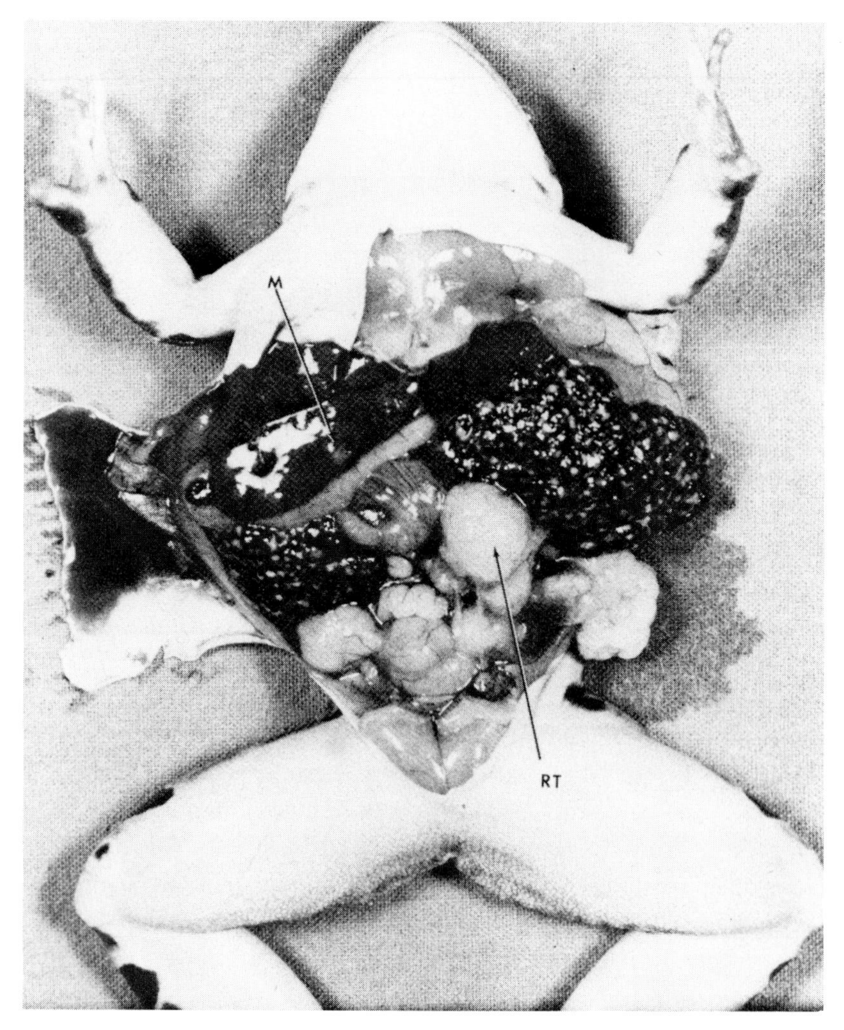

Figure 16. An autopsy of a mature leopard frog, *Rana pipiens*, reveals large lumps on the kidney (RT). These masses are a kidney cancer that has manifested its malignancy by spreading to the liver as a metastatic nodule (M). (From R. G. McKinnell, L. M. Steven, Jr., and D. D. Labat, "Frog renal tumors are composed of stroma, vascular elements, and epithelial cells: What type nucleus programs for tadpoles with the cloning procedure?" In N. Müller-Bérat, ed., *Progress in Differentiation Research*, © 1976 North Holland, Amsterdam)

chromosomes. It is now common knowledge that normal development demands normal chromosomes—abnormal sets of chromosomes in humans result in abnormalities such as Down's syndrome (mongolism). Therefore, the nuclear transplanter seeks cells with a normal chromosome constitution. The use of more sophisticated chromosome techniques enabled researchers to confirm Lucké's belief that the frog tumor was characterized by a normal number of chromosomes with normal form. Normal tadpoles and normal adult *Rana pipiens* have 26 chromosomes (Figure 17). The frog kidney cancer also has 26 chromosomes, and they appear similar or identical to those of the embryo and adult (Figure 18). The nuclear-transplantation procedure was perfected in *Rana pipiens*; nuclear donors should have a normal number of chromosomes; the frog cancer has a normal number of chromosomes—if any cancer was ideally suited for characterization by the cloning procedure, it was the frog kidney cancer.

I collaborated with the co-inventor of the nuclear-transplantation procedure, Thomas J. King, in studies of frog-tumor nuclear transplantation during the late 1950s. We found that instead of more cancer cells being formed when a cancer nucleus was placed in an enucleated egg, a tadpole was produced! That was an extraordinary observation in those early cloning days. Dogma said that cancer was an irreversible process. It was thought that a cancer cell could give rise only to more cancer cells. Presumably the control for the highly stabilized malignant condition rested in the nucleus, which contained the genetic material. Thus nuclear progeny of a cancer nucleus would be malignant. However, instead of giving rise to more cancer nuclei, the malignant nucleus produced progeny that contained cells differentiating as nerve, muscle, gut, etc. These results demonstrated that

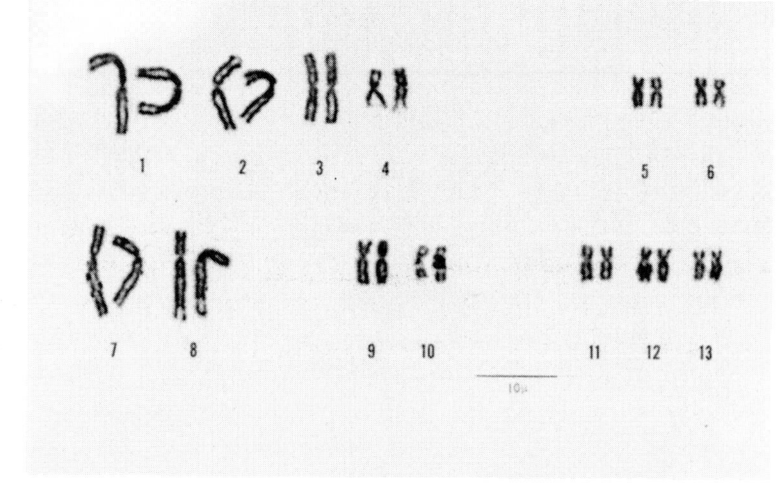

Figure 17. Chromosomes obtained from a normal adult leopard frog, *Rana pipiens.* (From DiBerardino, King, and McKinnell, 1963)

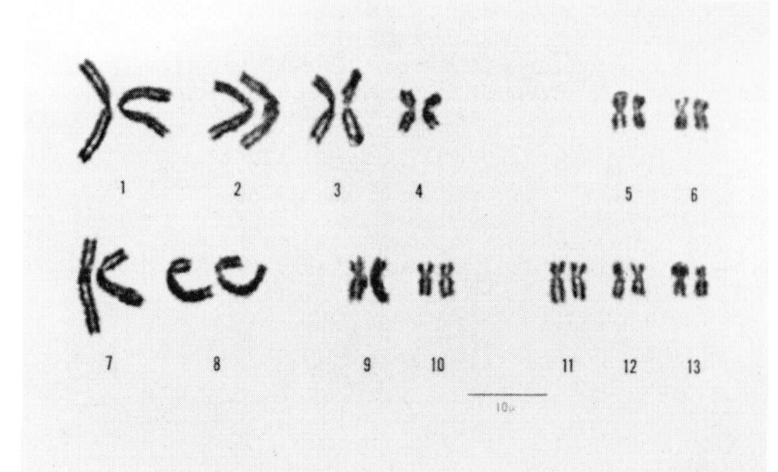

Figure 18. Chromosomes obtained from a frog kidney cancer. These chromosomes are indistinguishable in number and form from the chromosomes of a normal adult frog. (From DiBerardino, King, and McKinnell, 1963)

at the very least those genetic components of an adult cancer nucleus that were required for forming an early embryo were present and still capable of functioning. Further, and perhaps more important, the results demonstrated that cancer is not irreversible—at least frog kidney cancer is not.

The frog-tumor nuclear-transplantation studies were continued in Philadelphia by King and Marie A. DiBerardino and in my laboratory at Tulane University in New Orleans, and are continuing in my University of Minnesota laboratory.

At Tulane University, I produced tadpoles by inserting tumor nuclei into enucleated eggs (Figure 19). We were no less apprehensive about the possibility of parthenogenesis than other

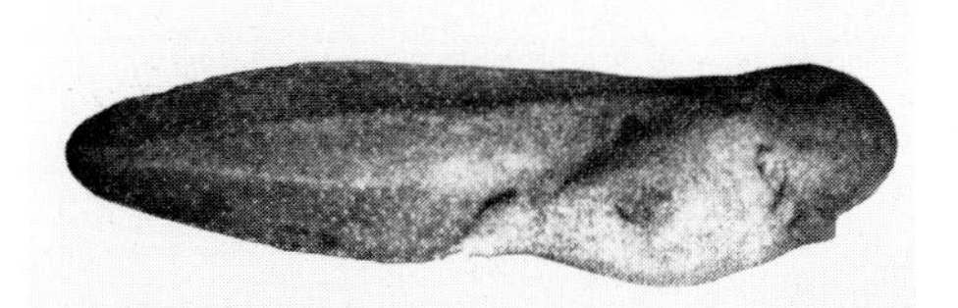

Figure 19. A tadpole produced by transplanting a kidney-cancer nucleus into an enucleated frog egg. (From R. G. McKinnell, "Nuclear transfer in *Xenopus* and *Rana* compared." In R. Harris, P. Allin, and D. Viza, eds., *Cell Differentiation* © 1972 Munksgaard International Publishers, Inc., Copenhagen, Denmark)

cloners had been. We wanted to eliminate the possibility that the tadpoles were produced from maternal egg chromosomes and to establish that the development of the operated egg resulted from the programming of a cancer-cell nucleus in enucleated cytoplasm.

I referred earlier to the problem of parthenogenesis in the context of the discussion of using donor nuclei tagged with mutant-pigment-pattern genes (Kandiyohi and Burnsi). The female providing eggs did not carry the mutant gene, and expression of the Kandiyohi or Burnsi characteristic thus constituted genetic proof of the donor nuclei's participation in development. The pigment-pattern-mutant genes Kandiyohi and Burnsi were not suitable nuclear markers for the cloned tumor nuclei because the mutants are expressed only in the adult state—Kandiyohi and Burnsi tadpoles do not appear to be different from ordinary tadpoles—and the tumor nuclear-transplant embryos died as young tadpoles.

A tumor nuclear marker—frog cancer with an extra set of chromosomes—was designed and produced in my laboratory in New Orleans and in the laboratory of Kenyon Tweedell of the University of Notre Dame. Although it is true that an abnormal complement of chromosomes is incompatible with normal development in frogs and humans, it has been known for some time that frogs and salamanders can develop normally (except for a reduction in fertility) with an extra *set* of chromosomes. The normal leopard frog has 26 chromosomes per cell, 13 derived from the sperm and 13 from the egg. An embryo with 23 or 24 chromosomes (i.e., 26−3 or 26−2) develops abnormally. But a leopard frog with 39 chromosomes (26 from the egg and 13 from the sperm) develops normally! The 39-chromosome frog is normal in every way except that it has reduced fertility (Figure 20).

A number of triploid embryos (embryos with three sets of chromosomes) were produced in New Orleans. They were produced by hydrostatic pressure which keeps a diploid egg diploid even after fertilization. As noted earlier, we wanted triploid tumors because the 13 extra chromosomes would serve admirably as a tag. Most parthenogenetic embryos either have one set of maternal chromosomes (haploid) or remain diploid, as the egg is

Figure 20. A triploid frog produced by the cloning procedure. (From McKinnell, 1964)

before fertilization. A triploid parthenogen would be an unlikely event. Thus triploid tumor nuclei should give rise to triploid embryos in a true nuclear-transplant operation. Haploid or diploid embryos resulting from the insertion of triploid nuclei would unquestionably be the result of parthenogenesis.

The triploid tadpoles produced in New Orleans were flown to Notre Dame, where Tweedell provided a cancer-causing virus preparation to inject into the tadpoles. The tadpoles were returned to New Orleans where they developed into frogs. Some had cancer, and the cancers were different from any naturally occurring frog malignancies. They were triploid (Figure 21).

Triploid tumor nuclei were then transplanted into enucleated eggs. Seven tadpoles were formed. The tadpoles swam. The seven swimming tadpoles had skin, connective tissue, muscle, brain, spinal cord, eyes, kidneys, liver, and all the other organs and tissues that characterize early tadpole anatomy. Swimming is significant because it demonstrates that not only are all of the requisite tissue types present and anatomically functional but

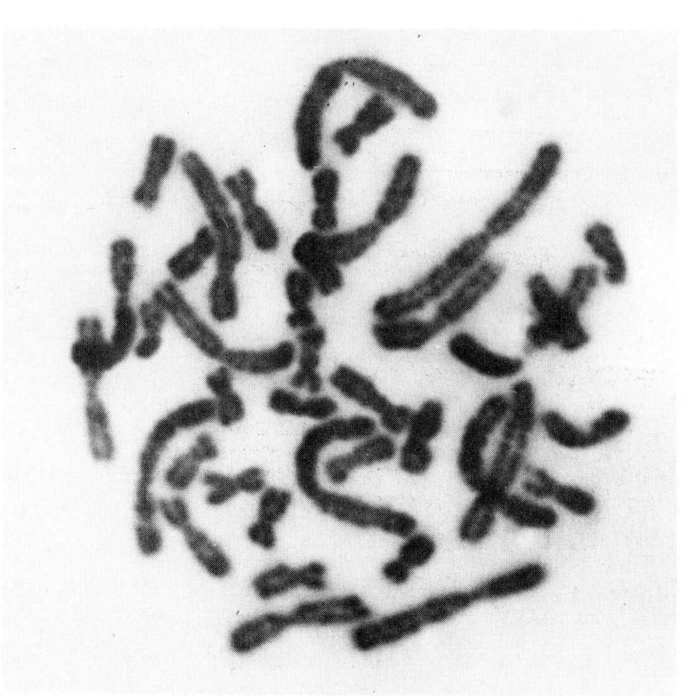

Figure 21. Chromosomes of a triploid kidney cancer of the northern leopard frog, *Rana pipiens*. (Photograph courtesy of Dr. Marie A. DiBerardino, from McKinnell and Tweedell, 1970)

the activity of the parts is coordinated. What was the chromosome number of these seven creatures? All seven were triploid. There was no reasonable possibility that the tadpoles could have developed from any kind of nuclei other than the nuclei obtained from the triploid *cancer*.

Renal cancers of frogs contain connective tissue as well as cancer cells. Therefore, it could be asked if a connective tissue nucleus was transplanted instead of a cancer nucleus in these studies. It is highly unlikely that any nucleus other than a cancer-cell nucleus provides genetic material for the tadpoles. Why do I believe this? My colleagues and I examined cells from frog renal tumors under a microscope with an ultra-violet light source. The cells had been treated with a chemical, acridine orange, that caused them to fluoresce when viewed with ultra-violet light. The microscopic procedure is similar to a procedure used to decorate some disco establishments. Black light (ultra-violet) shines on special paint and causes the paint to emit (fluoresce) bright colors of visible light. Cells can be treated so that they fluoresce, the glow revealing the *kind* of cell being studied. With fluorescence microscopy, we can distinguish between frog kidney-cancer cells and frog connective-tissue cells.

Several years ago, we wanted to ascertain what kind of cells are present in a dish following the dissociation procedure we use for tumor-cloning experiments. Did the dish contain mostly connective tissue cells, mostly cancer cells, or a mixture of the two? This is an important question because the kind of cell present in the dissociation dish is the kind of cell that provides a nucleus for cloning. We noted that cancer cells were present in excess of 98%. We knew they were malignant because the cytoplasm fluoresced red. The cytoplasm of noncancerous connective tissue from frog renal tumors fluoresces *green*. Because the overwhelming majority of donor cells were malignant, we could be confident that the tadpoles described above were in fact derived from malignant nuclear donors.

The genetic (triploid) evidence and the cell study (ultra-violet microscopy) evidence show that the hereditary material of a frog cancer cell contains determinants for the formation of many cell types other than malignant ones. The tadpoles from cancer nuclei also indicate that there is material in the egg cytoplasm that can coax or influence, in some as yet undefined way, malignant nuclei to give rise to nuclear progeny that include many seemingly *normal* types. If a frog had been produced in these cloning experiments, it would be evidence that the genetic material of a cancer cell is identical to the genetic material of a fertilized egg. Only tadpoles have been produced thus far with the cloning procedure using malignant nuclei.

The experiments raise questions: Do frogs not develop in cancer nuclear transfers because a virus is present? (Since the triploid tumors were formed by injection of a virus preparation, they presumably still contained virus.) Is the imperfect development attributable to the fact that the frog cancers were obtained from adults? The alternatives present a dilemma for which there is no answer at present. Recall, however, that the malignant nucleus evokes as much development of the enucleated egg cytoplasm as do normal *adult* nuclei which have been similarly studied.

Nuclear transplanters are interested not only in the capability of specialized nuclei to be reprogrammed, with the formation of a tadpole, but, perhaps more, in what it is within the cytoplasm that calls forth the reprogramming. Thus there are studies being done in a laboratory in Philadelphia on proteins that move into the nucleus in a nuclear transplant embryo, and vice versa. The bidirectional flow of protein, which may include extraordinarily important regulatory substances, is being studied there by Marie A. DiBerardino.

Earlier, the need for new kinds of chemotherapy was stressed —the need for nontoxic cancer-curing substances that affect differation rather than cause the death of cells. I would like to suggest that characterization of the proteins that move into and out of nuclei in the process of being converted from a malignant to a benign condition may be a first step in the development of a nontoxic chemotherapeutic agent.

IMMUNOLOGY AND CLONING

Heart transplantation, developed in the 1960s by several surgeons trained at the University of Minnesota, is now so advanced that it is not a failure in surgery that causes the untimely death of many heart recipients. Rather, death results when the recipient rejects the newly inserted heart. Although surgical skills are adequate for sophisticated tissue transplantation, the biology relating to rejection of foreign tissues is far less developed. It is apparent, therefore, that further knowledge of the immune response to tissue transplantation is urgently needed so that lives can be extended with the already developed surgical procedures.

Many animal species have been used in the study of how the immune response relates to tissue transplantation. Each species has its peculiar strengths and its distinctive weaknesses. Mice have been so useful in biomedical research that many strains with defined hereditary characteristics are available. Since tissue rejection has a hereditary component, mice are useful because of what we know of their hereditary makeup. However, tissue transplantation is often studied with individuals that do not reject grafts of tissues or organs—and with mice, this means that inbred individuals must be used. Inbred animals lack vigor, and are not similar to humans, which are the products of near-random mating. There are social, religious, and legal taboos that minimize inbreeding in humans.

We know less about frogs genetically than we do about mice. However, frogs resemble humans because of almost random breeding within a population, they are the most studied species of experimental vertebrate embryology, and cloning is best known in frogs. Because experimentalists have developed operative procedures, it is relatively easy to extirpate significant organs such as the thymus gland in a frog embryo—a significant gland because it is crucial in tissue rejection.

Can cloning be utilized in studies of tissue transplantation? To answer this question, the answers to two other questions are needed. First, do frogs respond to transplanted tissues with an immune reaction? Second, can frogs be obtained in groups that are as genetically similar as inbred mice but as vigorous as ordinary frogs?

The northern leopard frog responds to foreign tissue (tissue obtained from unrelated individuals within the same species) grafts in much the same manner as humans. Frog skin grafted to an unrelated frog of the same species initially heals and flourishes. However, after a time, the graft dies and is lost. The death and loss of the skin graft is due to an immune response from the host. The rejection is quick and vigorous. *Xenopus* also rejects grafted tissue, but the reaction is more sluggish and resembles that of lower vertebrates rather than humans.

Thus frogs have an immune response. But are there groups of genetically similar or identical individuals? Certainly not naturally, for natural populations of frogs result from random mating. However, the cloning procedure can produce groups of genetically identical individuals.

Ordinary cell division gives rise to two precise replicates of the genetic material. When the fertilized egg divides, the two blastomeres that are formed are genetically identical. Human twins derived from a common egg are examples. The identical twins look alike and can even receive transplanted kidneys from each other without fear of rejection. Subsequent cell division, like the first division, results in genetically identical nuclear progeny. Thus all the cells of a blastula (and presumably later stages, too, for that matter) are thought to be composed of identical DNA. Therefore if a number of frogs are produced by transferring nuclei from *one* blastula, the group should be as identical to each other as identical twins are.

I collaborated in a study with Peter Volpe of Tulane University in the mid-1960s, in which I produced several groups of frogs by cloning. A common blastula served as nuclear donor for each cloned group, making each group isogenic. Isogenic means that each individual within the group had the same genes as the other members of the group. Volpe transplanted embryonic and juvenile frog tissue among members of the isogenic group, and there was no rejection reaction (Figure 22). Tissue grafted between individuals of *different* groups was invariably rejected. The simple experiment was useful in demonstrating that genetically identical groups of animals, which were not inbred, could be produced and that grafts were not rejected within these groups.

In most nuclear transplantation experiments, there occur individuals with extra chromosome sets. They happen because the inserted nucleus divides before the egg cytoplasm divides. The two chromosome sets in one cytoplasm form two diploid nuclei. The diploid nuclei fuse, forming one nucleus that is double the

Figure 22. Cloned frogs that are genetically identical do not reject patches of pigment-containing cells grafted to their bellies. (From Volpe and McKinnell, 1966)

usual size and contains twice the ordinary amount of DNA. It is tetraploid. The tetraploid embryo becomes a tetraploid frog—since no new genes have been introduced in the isogenic groups, the tetraploid frog is genetically identical to the diploid frogs of the group. We already knew that the *quality* of genetic material affects the immune response; what we wanted to find out was whether the *quantity* of genetic material did so. The study of Volpe and McKinnell showed that tissue from a tetraploid individual was not rejected by isogenic diploid hosts and vice versa. This simple observation showed clearly that it is the diversity of genes, not the abundance, that is important in the rejection reaction.

Can cloning be useful in other immunobiological studies? Perhaps so, and one such study involves sexual histoincompatibility.

The genetic sex of a frog may be different from the appearance of the frog. A frog that looks and acts like a male frog may, in fact, have the genes of a female. Sex change is easy to accomplish in the laboratory by rearing tadpoles in water containing sex hormones. All tadpoles reared in a water solution of male hormone grow as male, even though about one-half of the tadpoles are genetically female. Similarly, all tadpoles reared in an aqueous solution of female hormone develop as females.

Female frogs reject male skin. Is rejection the result of genetic difference, the male having a chromosome the female has not, or hormonal incompatibility? The answers to these questions can be obtained by producing a large group of nuclear-transplant frogs from a common blastula, therefore isogenic. Half of the cloned embryos are reared in a male hormone solution, the other half in a female hormone solution. One individual is untreated to ascertain the true sex of the isogenic group. Do frogs that appear female and have the genes of a female reject tissue from frogs that appear male but have the same female genes as the graft recipient? I don't know the answer to that question yet. An associate is seeking it. I mention the problem to suggest how subtle immunobiological questions can be resolved with the cloning procedure.

Until immunobiology is understood sufficiently so that kidneys and hearts can be transplanted without a resulting rejection reaction, we will continue to need new information that can be used in the care of transplant patients. There is no question but that the cloning procedure can be used to obtain new information about immunobiology.

AGING—NEW INSIGHT THROUGH CLONING?

Aging may be the most important medical and human problem that can be served with the cloning procedure. As human populations become better able to contend with life-threatening medical problems such as infectious diseases, heart and kidney problems, and cancer, there is an increasing proportion of people who survive longer. At the same time, effective and widely used birth-control procedures result in fewer young people. It takes little imagination to contemplate the enormous social and economic problems of a reduced population of young people attempting to provide medical care for large numbers of incapacitated elderly.

Aging individuals who remain mentally alert and vigorously healthy support themselves at little or no cost to the rest of the population. In fact, they are a benefit to their community. They have time for the very young, they have vocational and professional experiences that can be a valuable resource to working people, and they may be reservoirs of community and family history. However, aging people who are mentally and physically incapacitated are an emotional and economic burden. The enormous cost of caring for these unwell, aged people will escalate. If not for humane reasons (*which are exceptionally compelling to me*), then for economic reasons we need to know more about the aging process now.

Some scientists are concerned with extending the life-span—my personal view is that enhancing the quality of life of mature citizens is more urgent. Actually, the discussion of extension versus quality may not be very important, for they may well go hand in hand.

The scientific understanding of aging in the 1970s may be compared to the biologists' perception of cancer in the 1950s. That is, two decades ago it seemed that if biologists would only bend their skills to the problem of cancer, new insights would emerge that would likely result in a better control of the disease. Although a cancer cure has yet to be developed, much of the work that has been accomplished in the last two decades supports the confidence that the goal of cancer-cell biology will yet be achieved.

Today, extraordinarily little is known about aging. There are virtually no theories that marry cell biology, genetics, and biochemistry to an understanding of aging. Perhaps in the next few decades biologists can provide new insight for the new science of how organisms, and humans, grow old.

A number of aging experiments with cells cultured in vitro have been performed by Leonard Hayflick, formerly of Stanford University and now at Children's Hospital Medical Center, Oakland, California. The experiments suggest that normal cells have the potential for only a limited number of replications before they die. Normal cells are characterized by normal chromosome complements. Hence, according to Hayflick, cells with normal chromosomes have a limited life expectancy in culture. However, not all cells in culture have normal chromosomes. Years ago a woman died of cervical cancer. Before her death, some of her cancerous cells were placed in a culture medium. The cells continued to grow after her death and are still growing. The cells seem immortal. How do they differ from "normal" cells in culture with a limited life span? Careful chromosome analysis revealed that the cultured cells derived from the cervical cancer contained an abnormal allotment of chromosomes. It would seem, therefore, that normal cells in culture have a limited life expectancy and only malignant cells persist. An attractive aspect of the studies of Hayflick is that because normal cultured cells can endure only a limited number of cell divisions before they die, the changes that occur in the culture before the demise can be characterized. Senescence refers to changes associated with aging. Are the changes in cell cultures with normal chromosomes the result of the senescence of the culture? Are the changes brought about because of continuous cell division? Is cell division *per se* a principal cause of aging?

A skeptic might observe that what transpires in a culture flask may have little to do with what happens in an inact organism. Why? The environment of cells in culture is designed by humans. It is not uncommon to find a mixture of fetal calf serum or embryo extract, plus amino acids, and buffers as a growth medium. It is remarkable that cells survive in such a milieu. Perhaps the demise of the culture after a number of cell generations is the result of growing in a highly artificial medium. Perhaps metabolites accumulate or the cells have a limited capability of surviving in the synthetic environment. The apprehensiveness of the skeptic is only partly allayed by cell culturists who point out that cells derived from young individuals have the capacity to undergo more cell divisions than do cells derived from old individuals. Is there an alternative and feasible mode of examining the effect of many cell divisions on aging? If there is, can the data derived from the alternative mode be exploited to provide new and useful information about aging? The answer to that question is yes—and at least one alternative mode of study is, of course, cloning.

The time required for a complete cell cycle, i.e., the time from one cell division to the next, is very short during the early development of *Rana pipiens*. When an embryo reaches the blastula stage, it has undergone 12 or 13 cell cycles and it is less than 1 day old. The duration of cell cycles increases as the embryo becomes older. As noted, a blastula nuclear donor has already undergone 12 or 13 cell cycles at the time it is inserted into enucleated egg cytoplasm. A nuclear-transplant blastula forms within a day after cloning and its cells have undergone 24 or 26 cell cycles. If the nuclear-transplant blastula serves as a nuclear donor, the donor nuclei will have sustained 24 or 26 cell cycles; when a blastula forms from this operation, the blastula cells will have experienced 36 to 39 cell cycles. The process of subcloning —also known as serial nuclear transplantation—can be continued day after day (Figure 23). Replication occurs with normal morphology of cells in a normal physiological environment. In theory, about 100 cell cycles could be induced by the end of day 8. It is known that serial cloning is feasible because it was first done, on a limited scale, by King and Briggs in 1956 and because we have already done pilot experiments in Minnesota.

Hayflick reported senescence in cell cultures after about 50 population doublings. Although the actual number of cell cycles is greater than 50 in such cultures (cells lost during culture-medium change are not counted), it is clear that it will not take an inordinate effort to look for cell senescence among serially cloned embryos.

How would aging be characterized in serially cloned embryos? I do not know because the experiments are only in the planning stage. Aging may not occur at all after serial nuclear transplantation. If aging fails to occur, it will be apparent that cell cycles *per se* are not responsible for senescence (at least in frogs). However, aging may be manifest in the serially cloned frogs in a reduced life span. Or it may be evident in an increased vulnerability to chromosomal aberrations, which may lead to a greater prevalence of abnormal development and tumor formation. There may be errors in DNA replication, or aging may be expressed as altered metabolism in the cloned frogs.

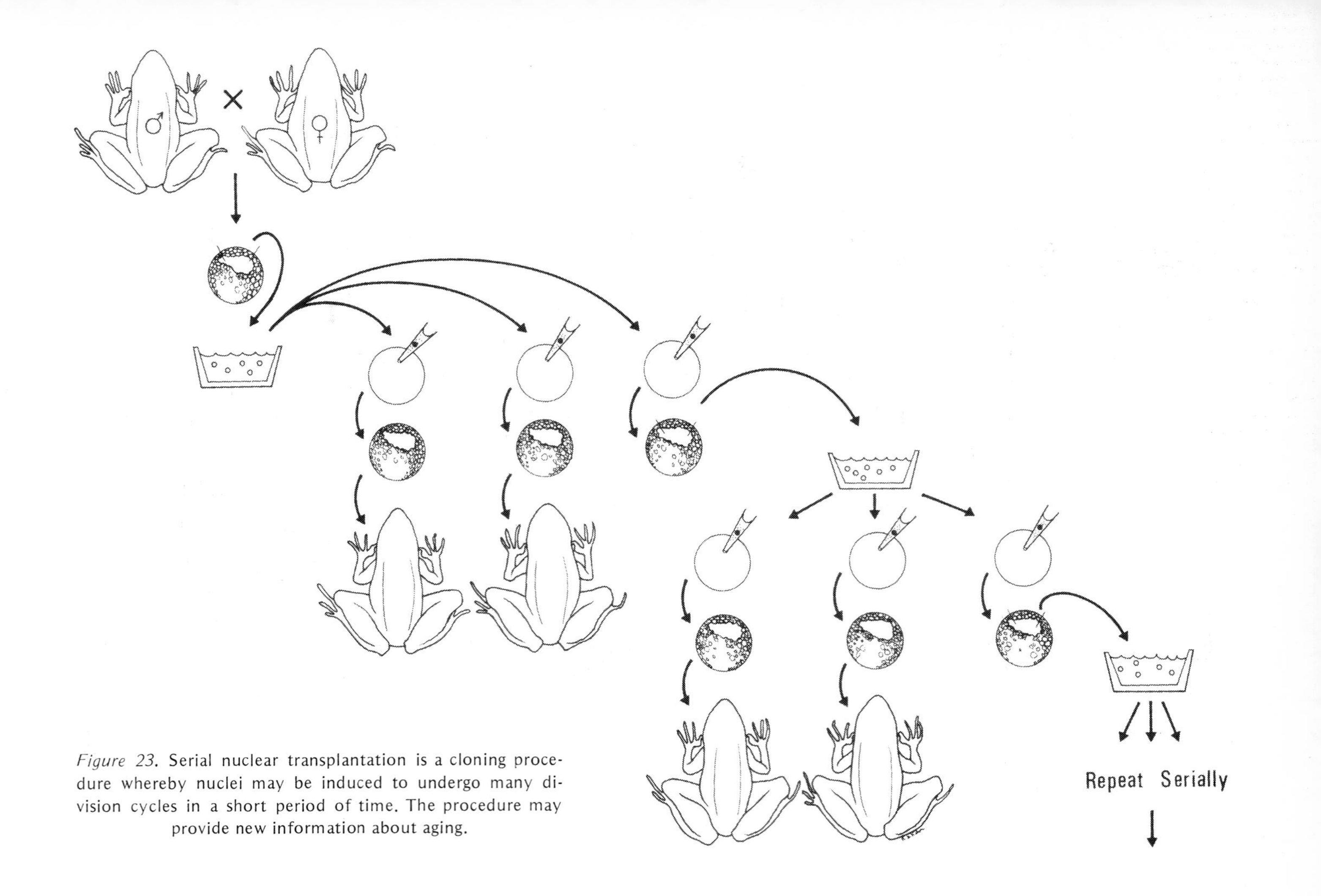

Figure 23. Serial nuclear transplantation is a cloning procedure whereby nuclei may be induced to undergo many division cycles in a short period of time. The procedure may provide new information about aging.

These comments are, of course, speculative. However, it is *not* conjecture to state that new information about the cell biology of aging can be generated with the cloning procedure. It is the business of the cloner, then, to use that information in conjunction with prospective studies to develop an understanding of aging that will benefit humankind.

Cancer, Aging, and Other Challenges

5

On Cloning Mice and Men

Although I am delighted to write about nuclear transplantation in frogs and other amphibia, I recognize that much of the general reader's interest is in the cloning of humans. I also perceive that most readers at the time of this writing understand that successful cloning of a human has *not* been accomplished. Frog clones are faits accompli, but thus far there is not even a cloned white mouse in existence, much less a human. Why? Is the mammalian egg too small for micromanipulation, as one Nobel laureate recently stated on nationwide television in the United States? Is the egg of a mouse or a rabbit too fragile for microsurgery? Is the reproductive biology of mammals so inadequately understood that it is too soon to contemplate cloning experiments in mice, rabbits, or humans? I believe the answer to these questions is "No."

I write this chapter to suggest that most, if not all, of the procedures for mammalian cloning are at hand. The tools and the experimental procedures seem to be here—quite frankly, I don't know why a mouse has not been cloned. But since this has *not* been done, despite the presence of what seem like reasonable procedures, I see little virtue in yielding to the temptation of speculating about *when* a mouse (or human) will be cloned.

On Cloning Mice and Men

What is needed for mammalian cloning? A source of eggs, an enucleation procedure, a supply of donor cells, a technique to put egg and donor nucleus together, and a means of culturing the developing clone until it can survive on its own are the essential prerequisites.

Before proceeding, let me comment on why I use the word mammals—rather than, say, humans—in this chapter. There was only a brief interval between the cloning of the North American leopard frog and the cloning of the South African clawed frog. Almost all frog eggs are the same size. If you master cloning in one frog species, you can without much difficulty extend that art to most other frog species. Since nearly all mammal eggs are the same size (the primitive platypus, spiny echidna, and marsupials such as the opossum being exceptions), the cloning procedure ought to be applicable to most mammals, including humans.

However, it is not entirely clear *which* mammal would be the optimal choice for the initial development of the cloning procedure. Mice are cheap and their genetics and reproductive biology are well known. However, each of their front feet has only four toes. Shrews are common, tiny (thus requiring minimal laboratory space), they have five digits on all four appendages, in contrast to mice but similar to humans, but some have a nasty bite. Rabbits have somewhat larger eggs than mice. Elephants are too big. Primates because of their obvious similarity to humans would be interesting, but primates are becoming less available as many countries become aware that their simian populations are being depleted by exports to the United States. Cloned humans would be newsworthy; but eggs in abundance would be difficult to obtain, and there are ethical and legal constraints relating to experimentation with humans. I think successful cloning will occur first in a small mammal such as a mouse or rabbit. Thus most of the discussion in this chapter pertains to small mammals.

SOURCES OF MAMMALIAN EGGS

It is not as easy to induce egg release from the ovary (ovulation) in a mammal as it is in a frog. However, any mammalian species can probably be ovulated at will. An incomplete list of mammals that have been ovulated by hormone injection includes rabbits, monkeys, voles, chinchillas, lions, squirrels, rats, shrews, mice, and humans. One procedure for inducing the release of the egg is the injection of a hormone known as pregnant mare's serum gonadotropin (PMS) followed in about two days with an injection of human chorionic gonadotropin (HCG). Curiously, the ovulation procedure works even better with immature mice than it does with sexually mature mice. The ova from the prepubescent mice are as capable of normal development as those obtained from mature females. Rabbits ovulate in response to the stimulation of sexual intercourse. But eggs can be obtained outside the mating procedure by hormone injection.

Eggs are, in theory, released from the ovary to the body cavity. In fact, however, most eggs move almost directly to the Fallopian tubes, which lead from the ovary to the uterus. It requires a bit of skill to surgically open a female mouse or rabbit, expose her reproductive organs aseptically, and wash out the ovulated eggs from the Fallopian tubes with warmed culture medium. The recovery of unfertilized eggs from humans is only slightly more difficult. Almost mature eggs (known as oocytes) can be plucked from mature ovarian follicles with a device that permits visualization and egg recovery via a very tiny abdominal incision. The device is known as a laparoscope and has been used extensively by Patrick Steptoe in England in studies designed to overcome infertility in women.

Freshly ovulated eggs are covered with a cloud of ovarian cells (Figure 24), known as cumulus cells, which must be removed with an enzyme to reveal the egg (Figure 25). A noncellular membrane, the zona pellucida, is found immediately under the cumulus cells; it ordinarily surrounds and protects the egg as it makes its passage from the ovary to the uterus via the Fallopian tubes. Since successful mammalian cloning has yet to be accomplished, I am unable to prognosticate whether or not it

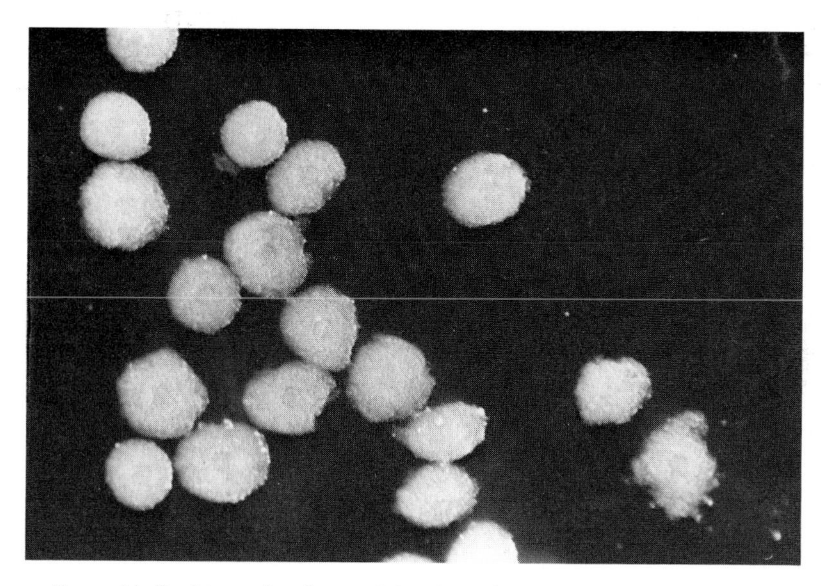

Figure 24. Freshly ovulated ova of the short-tailed shrew, *Blarina brevicauda*. The eggs are partially obscured by a cloud of cumulus cells. The shrews were wild-caught in Minnesota, ovulated in the laboratory, and their ova were photographed by S. Goustin, R. Stanek, and R. Werner, former undergraduate students of the author

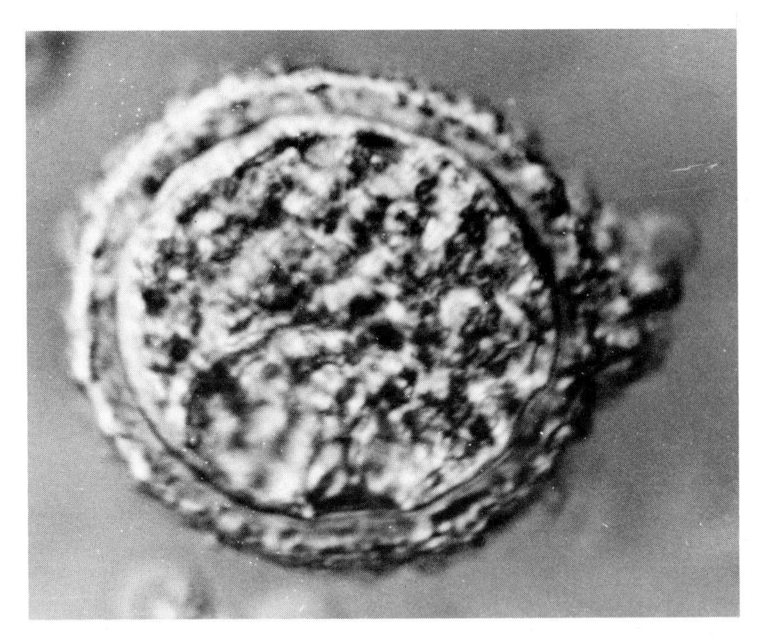

Figure 25. An ovum of the short-tailed shrew, *Blarina brevicauda*, with cumulus cells removed by enzyme treatment. Photograph by S. Goustin, R. Stanek, and R. Werner, former undergraduate students of the author

would be wise to remove the zona pellucida. In the experimental probes related to the development of a cloning procedure in mammals, some experimenters leave the zona intact and insert a microinstrument through the membrane (Figure 26). Others use an enzyme, pronase, to assist in the removal of the zona. Time will tell whether or not cloning in mammals will require the routine removal of the zona.

Enucleation

Frog eggs may be enucleated manually with a glass needle. Would a similar procedure work with a mouse or a shrew egg?

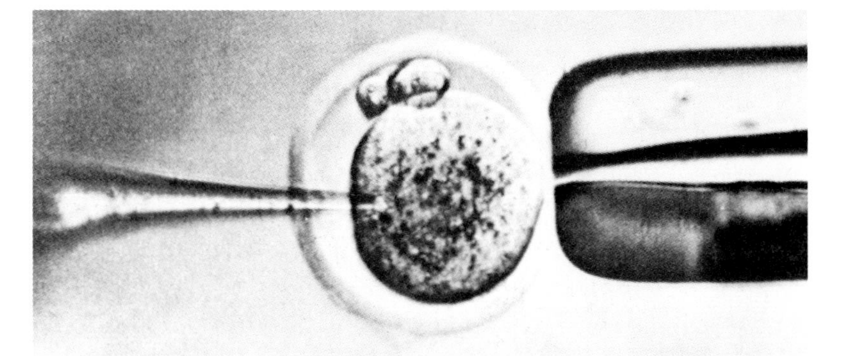

Figure 26. A mouse egg can be subjected to micromanipulation and microsurgery and survive. (From Teh Ping Lin, "Microinjection of mouse eggs," *Science*, 1966, 151:333-37, Copyright 1966 by the American Association for the Advancement of Science)

Probably not. Mouse and shrew eggs are about a tenth the diameter of a frog egg. A rent in the plasma membrane sufficiently large to permit the removal of maternal chromosomes might be so large that it would allow leakage of cytoplasm, thus precluding normal development. Lest the idea of surgical removal be abandoned entirely, I will note that McClendon, mentioned previously in the discussion of the development of isolated frog blastomeres, enucleated *Asterias* (a marine starfish) eggs by sucking out the chromosomes (Figure 27). This was accomplished on eggs about the same size as mammalian eggs—and it was reported in 1908. Thus surgical enucleation is not precluded in principle because of egg size. Recently one of the two nuclei present in a mouse egg shortly after contact with sperm was successfully removed by cell surgery. The survival of the operated egg, which has only a female genetic parent, is evidence of the capacity of a mammalian egg to survive surgery.

Although the removal of egg chromosomes by surgery is a possibility, I think that enucleation by radiation will probably

Figure 27. Surgical removal of the nucleus of a starfish egg. (McClendon, 1908)

be less traumatic for the egg. Laser, ionizing, and ultra-violet radiation all hold promise as potentially feasible enucleation methods.

Would laser irradiation be an effective means of removing maternal genetic material? I have used a ruby laser to enucleate frog eggs. It works well if the irradiated object is pigmented. The mouse egg, of course, does not contain pigment (melanin) granules. However, individual mammalian blastomeres can be ablated with laser irradiation after the cleaving ovum has been stained with a dye that does not kill the surviving blastomere. Precision laser irradiation technology is developing rapidly, and parts of a single chromosome can be irradiated. Thus it is premature to conclude that the laser would be an inappropriate means of enucleating mammalian eggs.

Hajima Sambuichi, formerly of Hiroshima University and currently at Yamaguchi University, showed that frog-egg cytoplasm is not harmed, within limits, by enormous doses of gamma irradiation. Normal nuclei inserted into heavily irradiated cytoplasm result in some cloned frogs that develop to metamorphosis. Since ionizing radiation (such as gamma or X-rays) affects nuclei and seems not to damage cytoplasm, mammalian eggs perhaps will be enucleated in this manner. Ultra-violet radiation is used to enucleate the eggs of *Xenopus* before nuclear transplantation. Methodology may be developed for the utilization of ultra-violet in mouse experiments.

"Chemical enucleation" is a new term. One meaning is using chemical agents to extrude a nucleus. Cytochalasin B and colchicine are both naturally occurring chemical substances which, under appropriate circumstances, will cause nuclear extrusion in mammalian cells. If these, or similarly acting, chemical agents are used for enucleation, precise doses and mode of administration will have to be worked out because the drugs are potentially very toxic.

Although there are several methods available for enucleation, none as yet has emerged as the method of choice. In that respect, the early frog cloners had a substantial advantage over the mammalian embryologists. Keith Porter developed an adequate enucleation procedure for frogs in 1939, and the procedure was 13 years old by the time the first frog cloning experiments were reported.

NUCLEAR DONORS

The first successful cloning in mammals will most likely be of embryonic nuclei. This prediction is, of course, based on extrapolation from amphibian studies. Recall that gastrula and older nuclei of frogs do not perform in egg cytoplasm as well as blastula nuclei of frogs. Thus a supply of precisely aged young mammalian embryos probably will be needed.

Two methods of obtaining early embryos are available: removal from the reproductive tract of the recently mated female and in vitro fertilization.

Fertilization in humans, mice, and rabbits ordinarily occurs in the Fallopian tubes. It takes several days for the young embryo to move through the tube and arrive at the uterus. Although it is true that implantation of the embryo occurs in the wall of the uterus and removal of the fragile embryo from the site of its attachment would be an exceptionally perilous maneuver, the experimental embryologist has substantial time available before implantation when free embryos can be flushed from the reproductive tract (just as unfertilized eggs can be washed from the tubes).

There is some merit in considering in vitro fertilization as a source of early embryos. Why? The precise moment that sperm makes contact with the egg can be recorded. The development of the fertilized egg can be documented. In vitro fertilization of mouse, rabbit, hamster, and human ova has been reported in the scientific literature. Mouse ova, fertilized in the laboratory, develop normally when reimplanted into receptive female mice. A girl baby was born recently in England and is reportedly doing well. The baby started life as an egg fertilized in the laboratory. In vitro fertilization involves the recovery of freshly ovulated eggs and exposure to sperm that is competent to fertilize. Curiously, sperm from a normal ejaculation is not necessarily competent to fertilize eggs in vitro. Sperm obtained from the coiled tube just above the testis (epididymis) will fertilize mouse eggs. Sperm in fresh human semen must be washed before it is in a physiological state that will permit fertilization in the laboratory.

Fertilized eggs divide and give rise to smaller and smaller embryonic cells. If one wishes to transplant the nucleus of a young embryo, one must first remove the embryo from the zona pellucida. The zona pellucida may be digested with an appropriate enzyme to expose the embryo for cellular dissociation. The embryo should be protected from the harsh enzyme by removing it from the solution before the zona is completely digested. The delicate remains of the zona can then be removed with very fine tweezers or by swirling the embryo in and out of a medicine dropper. Free, unattached cells can be obtained from the denuded embryo by exposure to another enzyme solution such as trypsin. The trypsin presumably digests the material that is responsible for cells sticking together.

Thus far, eggs and their enucleation and embryos and their dissociation have been described. Only two steps remain for the cloning of a mammal. One step is nuclear insertion and the other is the culture and care of the cloned embryo.

All procedures described thus far are subject to variation. There is no standard mode of cloning in mammals because they have yet to be cloned. For example, investigators may omit enzymatic digestion of the zona pellucida and instead rely upon mechanical cracking of the zona. Or they may dissociate the embryonic cells by bouncing the embryo on a mattress of agar until the cells fall apart. Cracking the zona and mattress bouncing were described in a scientific paper by J. D. Bromhall of Oxford University, England.

Micromanipulation to Join a Nucleus with an Enucleated Egg

Mammalian eggs have the remarkable capacity to survive surgery while being grasped in an egg-holding micropipette. Professor Teh Ping Lin of San Francisco described removal of egg cytoplasm with a micropipette and microinjection of solutions into an egg in a series of papers published in the 1960s. Surely, if a mouse egg can be manipulated in such a manner and survive, then it ought to withstand the shock of enucleation and the surgical insertion of a new diploid nucleus.

In fact, experiments have been published that suggest the mammalian egg has durability. It is not known for certain at this time if eggs will survive the cell surgery of nuclear transplantation so that a mouse, rabbit, or human will result from the operation. But certainly there are preliminary operations that stimulate the imagination—and, incidentally, send science-fiction writers back to their typewriters. Recently, Polish biologist Jacek A. Modlinski, of the University of Warsaw, published his method for inserting embryonic nuclei in *fertilized* mouse eggs. Remember that cloning requires that the host egg be enucleated. Therefore, the Polish experiments are not true cloning. But they do illustrate elegantly that a mammalian egg can be grasped with a polished, blunt micropipette; an embryonic cell can be sucked into a sharpened micropipette; and the nucleus from the embryonic cell can be deposited in the cytoplasm of the host egg. Modlinski reported that the operated egg survived for a relatively short time (five days) but in only a very low percentage of cases. The transplanted nuclei contained "marker" chromosomes, and the chromosomes were identified in four of the nuclear transplants. The embryos contained double the normal dose of chromosomes—one set from the donor embryo, the other set from the fertilized egg. Thus it is abundantly clear that embryonic nuclei inserted into egg cytoplasm can survive the trauma of transplantation for at least a few days.

More recently, Landrum B. Shettles reported in the *American*

Journal of Obstetrics and Gynecology (January 15, 1979) that he transplanted a human nucleus, which resulted in the formation of an early embryo. The transplanted nucleus was derived from a human spermatogonial cell (diploid precursors of the mature haploid sperm). The human egg recipient was enucleated with a micropipette. Several operations were performed, three resulting in ova that formed small clusters of cells which resembled mulberries (morulae). The three human morulae were then discarded. Shettles suggested that normal development would have resulted had the morulae been inserted in the uteruses of humans.

What is the biological significance of the mulberry-like cluster of cells? It is impossible with the information presented in the article to answer that question. I think it highly unlikely that the cells would have developed to the adult stage. I base this opinion on the results of analogous nuclear-transplantation experiments performed on nonhuman subjects.

Marie DiBerardino and Nancy Hoffner inserted frog spermatogonial nuclei into the cytoplasm of enucleated frog eggs in 1971. Abnormal larvae were produced, the longest-lived surviving 20 days. *None developed into frogs*. The DiBerardino and Hoffner results are entirely in harmony with results of nonhuman adult nuclear-transplantation experiments reported from other laboratories. That is, nuclei from adult cells do not seem to be competent to promote normal development to the adult stage. If the morulae in Shettles's experiment had the capacity to develop normally when inserted in human uteruses, they were truly exceptional clones.

A second reason for skepticism about the success of the Shettles experiment relates to his failure to use a nuclear marker. As I said earlier, it is necessary in cloning experiments to use markers to show that development is programmed by the inserted nucleus, because there are several other conditions in eggs that could be confused with successful cloning. One such condition is fragmentation. For example, in one study, more than three-fourths of the unfertilized pig eggs obtained from a sow's uterus within two days of ovulation fragmented—and some of the fragmented eggs resembled *normal cleavage stages*. Despite their normal appearance, fragmented eggs are in an early stage of degeneration. Were the discarded cloned morulae in the Shettles experiment undergoing fragmentation which resembled cleavage? Another egg condition relates to abnormal chromosomes. Nonhuman embryos cloned from nuclei with abnormal chromosomes cleave normally but fail to develop properly at a later embryonic stage and generally die. Did the morulae have normal chromosomes? Finally, I remind the reader of the starfish ova of McClendon, the sea urchin ova of Ethyl Browne Harvey, and the frog eggs of Briggs, Green, and King—all of which had the capacity to divide with *no nucleus at all*. Of course, no development beyond cleavage was possible in the nucleusless eggs.

For these reasons, I think Shettles's suggestion that normal development would have followed the insertion of the experimental morulae into the uteruses of humans is at most speculation.

Uniting Egg Cytoplasm and Donor Cell Nucleus

Micromanipulation, although conceptually simple and mechanically straightforward, is not the only way to bring together a somatic cell and an egg. Cloning can learn lessons from cell culture, and some of the lessons relate to nuclear insertion. Body cells of various kinds have been grown in glass vessels for years. What happens if cells of two types are cultured together? Somewhat less than two decades ago, spontaneous fusion of two cell types was observed. Joining disparate cell types was not a common event. However, the frequency could be increased by using a virus. As a parenthetical note, it may be observed that a cloner should have a working knowledge of virology, cell culture, as well as mammalian reproductive biology. The virus that is useful in increasing the rate of cells joining together as one cell is known as Sendai. The Sendai virus increases cell fusion many hundreds of times, and it is not surprising that it has the disadvantage of being toxic to culture cells. However, careful irradiation of the Sendai virus with ultra-violet light, or treatment of the virus with certain chemicals, will alter its virulence without substantial diminution of its capacity to induce cell fusion.

Would it be possible to induce a somatic cell to join with an enucleated egg with the help of the Sendai virus? Yes. Certain preliminary studies suggest that possibility. Christopher Graham of Oxford University, who works with mice, fused mouse spleen cells with unfertilized and fertilized mouse eggs with the aid of the Sendai virus. The experiment was more significant as a model of what may come than for its results. Graham reported that cleavage of the treated egg was abortive or blocked at the 2-cell stage. As in the Modlinski study, the egg was not enucleated.

J. D. Bromhall, mentioned previously in regard to bouncing embryo cells on an agar mattress, has effected nuclear insertion in rabbit eggs both by surgical means and by virus fusion. Such eggs can occasionally undergo cell division until a morula is formed.

The experimenters cited above do not include all who have made significant contributions. However, their results are typical of those currently being obtained with nuclear insertion. The results are sufficiently encouraging to urge continued attention to the development of technique. However, the results obtained thus far do *not* permit the claim to be made of successful mammalian cloning. This is still beyond the capability of experimental embryologists.

Husbandry of a Cloned Mammal

The preceding suggests that it may be possible, sometime in the future, to assemble a developmental system composed of an enucleated mouse or rabbit egg and a competent nucleus from a body cell. The mammalian embryo, produced under these circumstances, must then be cultured if one is to obtain a mouse or a rabbit. In theory, the embryo could be cared for in vitro or in the uterus of a suitable female. In vitro husbandry of mammalian embryos is not the simple task of amphibian embryo culture. Most children have cultivated tadpoles by placing them in a dish of water with food, but no scientist has ever cultured a mammalian embryo in laboratory glassware beyond the earliest embryonic stages. Since in vitro care to birth has *not* been accomplished with any mammal, it is appropriate to examine what has proved feasible—foster motherhood.

In the style of the social commentator who suggested that birth control pills make sex without reproduction possible and who also suggested that cloning makes reproduction without sex possible, may I suggest that the biological technique of foster motherhood (also known as surrogate motherhood) makes pregnancy without sex *or* parenthood possible.

Foster motherhood in mice, that is, the placing of fertilized eggs or early embryos in the Fallopian tubes or in the uterus proper of a female has been possible for several decades. Skeptics may wonder how a biologist knows that a new-born mouse is *not* the biological offspring of a particular female. Genetic markers, such as black coat color, of the implanted embryo will distinguish it from the natural progeny of a white mouse. The use of females as incubators and caretakers for offspring that are not their own is not limited to mice. Cows have served as hosts to developing calves that were not their own progeny. And it seems that it would be possible in humans, because a baby was born in England, the child of its own parents, that was introduced into the uterus of its mother after fertilization in the laboratory. There is no reason to believe that foster motherhood involving a normal cloned mammalian embryo would not be successful—that the offspring would not be carried to term and eventually be reared to sexual maturity.

Cloning of Farm Animals

Thus far, my comments about mammalian cloning have been biased because of my experience with frogs. The bias relates to the rationale for cloning. I have emphasized in this book that reproduction is not the motive that drives frog cloners—frogs are cloned to provide new information for biology and medicine. However, there may be agricultural applications for cloning, if the procedure becomes practical, and these applications relate to reproduction. I referred earlier to the desirability of uniformity and high quality in plants which has led to cloning in horticulture. A similar need may result in the use of cloning in animal production.

Cattle (or pigs or sheep or other farm animals) produced by ordinary sexual reproduction vary. Because of this variation, some animals are more desirable than others. Unfortunately, it may take four or five years for a particular farm animal to reach sexual maturity, and even though it may prove to be particularly valuable (because of outstanding milk production, quality of wool, etc.), the progeny of that exceptional animal, produced by sexual reproduction, vary. Enhancement of animals by sexual selection is a slow process. It would be far more efficient to clone the tested and proved animal than to rear highly variable animals produced by sexual reproduction. Thus were mammalian cloning possible, it would surely be used in farm-animal production. Body-cell nuclei of proven exceptional animals would be inserted into enucleated host eggs of ordinary females. The cloned farm animal would have all the desired genetic characteristics of the nuclear donor animal. No effort or feed would be wasted on genetically inferior animals.

Agricultural scientists would probably also use cloning for studying variables relating to important characteristics like milk production, feeding regimens, and disease resistance. These studies are expensive since many animals must be used because of their genetic diversity. If isogenic animals (genetic replicates), produced by cloning, were available, fewer animals could be used in these experiments. The savings would be considerable.

When will mammalian cloning occur and in what species? It seems foolish to attempt any prediction other than that it will eventually be done. To believe that cloning cannot be accomplished in mammals is to suggest that mammals differ in some fundamental way in genetics, cell biology, and biochemistry from other vertebrates. The thrust of modern cell biology is that cells are cells, chromosomes are chromosomes, enzymes are enzymes. Thus I am confident that mammalian cloning will occur—eventually.

6

A Hundred Einsteins?

An elephant has not yet been cloned. Neither has a shrew, a mouse, a rabbit, or a human (successfully). Although little has been inscribed about cloned shrews and other small mammals, much has been written about clonal humans. Most is the work of science-fiction writers, philosophers, theologians, and lawyers. As far as I know, biologists experienced in cloning technology have *not* written about any aspect of cloning other than their own experiments. Perhaps I am an exception because I, as a teacher in a university classroom, respond to frequent questions about the implications of cloning humans. It seems appropriate that a nuclear transplanter contemplate the significance of his discipline and be willing to discuss the subject for the benefit of students and other interested individuals. That is the rationale for the first part of this book. Obviously, speculation is just that regardless of the qualifications of the speculator. I wish to assure the reader that much of what follows is speculation, that it represents my thoughts and does not necessarily represent the views of other cloners.

If cloning were available for the benefit of humans, how would it be used? Clonal astronauts, armies of genetic replicates, or spare parts? These concepts are not mine, but I consider them here because they have been written about so much and because they suggest some biological problems that can very properly be addressed by a biologist, especially one with experience in cloning.

COSMONAUT CLONATES—BRILLIANT BUT HOW BENEFICIAL?

Consider the possibilities. Why tolerate a hodgepodge of nearly middle-aged people of variable stature, intellectual capacity, and emotional stability on an expedition to the moon? Why not find *the* man or woman with psyche and soma that is the quintessence of all that is noble—and send clones of so splendid an individual to the moon. The idea is appealing. It weds late twentieth-century rocket engineering to late twentieth-century biological concepts in a conjugal fantasy of artificial humans and real but hard-to-believe rockets.

However appealing the idea, the postulated lunar clones are in some ways inadequate. The deficiencies of the group relate to its identicalness, including chronological age. If such a group were produced, it would take no less than 25 to 35 years for the clone to be of appropriate intellectual development and have the requisite manual skills to run a spaceship. A cloned human will mature no faster than an ordinary human, just as it takes no less time for a cloned frog to grow than it does for an ordinary frog. I suspect that the pace of scientific and technological development will increase, which will make prognostication of future needs exceedingly difficult—and delay will not be easily suffered. Selecting a genotype for a lunar or planetary mission 30 years in the future would be like having tapped John Glenn in 1931 for his 1962 global orbit in *Friendship 7*. Few can plan so far ahead.

There are other potential problems relating to chronology. All conventional clones of replicate astronauts would be the same age. Is this desirable? Many believe that intelligence does not decline as humans age. Contrary to what was previously supposed, intelligence as measured in at least some tests actually increases into the sixth and seventh decades. The genes do not change over the decades. What changes is the expression of the genes, and this is what intelligence tests measure. Judges, philosophers, and statesmen frequently blossom in their sixth or later decades. Ice skaters, molecular biologists, gymnasts, and military jet pilots flower younger. Would we really want a space capsule inhabited entirely by individuals of identical age? Space probes in the 1970s were populated by men of different ages.

Certainly, biocrats with the resources to produce clones of space explorers would also have the resources to produce a group of genetically identical but chronologically disparate, splendid space explorers. Genetic replicas need not be the same age. If we envision the consequences of cloning humans, we may envision another biotechnical achievement of the twentieth century—cryobiology. Human cells survive for at least 16 years at −78°C. Intact embryos of mice may be frozen, thawed, and transferred to foster mothers, with normal live births ensuing. Thus our procedure for producing genetically identical individuals of different ages would be as follows: identify individuals with appropriate genes and put their body cells in culture; freeze the culture; every several years thaw a sample of frozen cells and clone them. An alternative to this procedure would be to clone a number of individuals at one time, freeze the embryo clones at an early stage, and then at intervals of several years, thaw and transfer the cloned embryo to a receptive foster mother who would give birth to the future astronaut 9 months later. Thus we would obtain a group in which we exploit a superb genotype during the strength and vigor of its youth and simultaneous with the wisdom and judgment of its later years.

Feminists should resent an all-male clone group. There are intrinsic differences between males and females. An example: although females generally have less brute strength, they survive longer. Let the unisexuality of conventional clones not interfere with the desire to have a truly magnificent isogenic group.

The biologist who has muscle enough to manipulate genes and ages would also have the technical strength to manipulate the expression of sexual characteristics. For many years it has been possible to control sex among vertebrate embryos by adding the appropriate hormone. Why not manipulate the expression of sex-determining genes in astronaut clones? Deliberate sex reversal (in contrast to sex-surgery alteration) has not been accomplished in humans. However, human cloning has the prerequisite of egg enucleation and successful cell surgery. Given this as yet to be accomplished capability with human cells, it seems likely that the simpler technique of early administration of estrogenic hormones in vitro would provide anatomically appearing females from genetic males. Similarly, testosterone in a culture medium would result in females that were anatomically male. Thus a genetically male clone would be manipulated with hormones so that both male and female individuals would ensue. A clone of lunar explorers, *not* like those envisioned by science writers, but of various ages and both sexes, and all of a single glorious genetic constitution, could be engineered.

Two considerations become obvious at this point. One relates to the rather formidable biotechnology that has developed in recent years. The discussion of proposed and seemingly feasible variant clones witnesses to the impressive control over reproduction, growth, and differentiation that is possible in the late twentieth century. The other relates to the diversity in a microcosm (such as a space capsule) and diversity in this goodly frame, the earth. Most of us will agree that a diversity of kinds of people is desirable in communities such as universities, factories, armies, and nations. How do you run a university with only art historians? Or newspapers with editors but no reporters or printers? A spaceship lacking diversity among its personnel would probably be unsafe and dull. If diversity is required, why not obtain it in the conventional way, by finding qualified individuals produced, perhaps prosaically but nevertheless usefully, by sexual reproduction. This guarantees diversity far in excess of that which results from manipulating factors of isogenic groups.

So, the basic question is: Are clonal cosmonauts necessary—or even desirable?

THE MENACE OF AN ARMY OF GENETIC REPLICATES

Détente results in easing of fears about what occurs behind an iron, bamboo, or plastic curtain. Tension grows when détente falters—and fears return. It has been suggested in the nonscientific literature that some distant enemy may develop a nascent army of powerful, unfeeling human automatons—produced by cloning. They are a potential menace presumably because they are not the results of conventional sexual reproduction. They are pictured as unfeeling monsters. Surely such an army would move across a battlefield in a way so terrible that it almost defies the human imagination. Do we compete by producing our own cloned armies, or do we bow to extinction? Before this question causes the reader apprehensiveness and consternation, let us recall the previous discussion on cloned astronauts.

Lives there a dictator who would create an army to do battle 18 years hence? Military and political leaders thrive on power—not latent but highly visible and immediate power. Both Adolph Hitler and Franklin Roosevelt, men of enormous authority and command, assumed political control of their respective nations in the 1930s. Had cloning been available to each at the onset of his career as a national leader, no political consequence would have ensued because both were dead when their hypothetical cloned armies would be entering pubescence.

I suspect that a despot wants more from biotechnology than an isogenic army, with unproved characteristics, that will not be available for almost two decades.

100 MR. EINSTEINS AND RECONSTITUTING SLAIN LEADERS

Genetically identical Albert Einsteins and Adolph Hitlers have been illustrated in popular journals. Several years ago Nobel laureate Joshua Lederburg wondered what the characteristics of a cloned version of Nobel laureate Einstein would be. If one assumes that a cloned Einstein would have at least some attributes of Einstein, it is not unreasonable to assume that a cloned political chieftain would have some of the characteristics of that chieftain.

But *how much* would replicate Einsteins, Hitlers, or future dictators resemble their originals? I don't think the carbon copy would be a faithful replica of the prototype. Consider the social scene in the nineteenth century Europe of Einstein (born in 1879). Contrast that with our world. If someone had cultured body cells from Einstein before his death, had frozen them, and then transplanted his thawed, body-cell nuclei in the 1970s, no notion of the outcome of the cloning experiment would be available until the early 1990s—over a century after the original was born! The cloned Einstein would miss World War I and World War II. Wouldn't this make the clone in some ways different from its original?

I first crossed the Atlantic Ocean on a ship that took six days from Quebec City to Liverpool. I was in my twenties. My youngest child, with her siblings, crossed the Atlantic in six hours during her first year of life and crossed it twice again by age six. I am convinced that the ease of travel, the instant communication in our world, and other late 20th-century developments, will expand the intellectual horizons of my children. They cannot be the same children they would have been had they been born when I was. How could a 20th-century Einstein be the same as a 19th-century Einstein? The replicate would not even look the same as the original. Growth patterns have changed in the past half century. It is well known that soldiers in World War II could not fit into the uniforms their fathers wore in World War I. Although Einstein's genes might remain stable, the *expression* of his genes in replicate Einsteins certainly would not.

WHERE EXPECTATION EXCEEDS CAPABILITY

The capacity of biologists to fabricate genetic replicate frogs has led to an anticipation of carbon-copy humans. It is perhaps in this arena of biological endeavor that the capacity (or will) to produce and the popular expectation diverge the most. I believe it is rational to expect cancer control in the future; I believe it is sensible to anticipate new insights into the immune process; I believe that graceful aging is a reasonable hope for a greater proportion of our population in the future. I have faith that cloning will contribute to each of these areas. However, I never expect to witness the construction of carbon-copy humans. I do not believe that nuclear transplantation for the purpose of producing human beings will ever routinely occur.

There are two compelling reasons for my conviction. The first, which I discussed earlier, relates to the indispensable biological requirement for diversity—diversity that is most efficiently gained through conventional sexual reproduction. The second reason, discussed in the previous section, relates to the ability of the cloning procedure to produce carbon copies. Genetic replicates perhaps—carbon copies never. Epigenesis is a term biologists use to describe the unfolding of potentialities in a developmental system. A fertilized egg of a sea urchin or a frog is not a miniature sea creature or a tiny hopping amphibian. Rather, it is a single cell that has, under certain environmental conditions, the capacity to develop into an adult organism. The fulfillment of that capacity is absolutely wedded to the environment. If the reader does not believe that, try placing a marine egg in hot water or permit a frog-breeding pond to dry up. Expectation is not fulfilled in the modified environment. Development is contingent upon an environment. Epigenesis is plastic in that minor fluctuations in the physical environment result in minor fluctuations in the structure of an organism. A slight diminution of an essential growth factor in an experimental animal's diet results in somewhat less growth.

The social environment of the human organism is far more

complex and subtle than the physical environment. Personality is as epigenetic as is the soma. The zygote of Hitler was not a despotic leader. Neither was the neonate Hitler a dictator. Circumstances—history—the environment—evoked the genes of young Adolph Hitler to be expressed as the oppressor we remember. Although it is true that biologists can induce the replication of a genetic apparatus in certain instances, they do not have now nor are they ever likely to have the power to clone the social environment. The feature of development known as epigenesis prevents the creation of carbon-copy people. Thus expectation exceeds capability—even in the late twentieth century.

A boy baby reared as a girl baby acts like a girl baby. A contemporary American acts like a contemporary American. Had Mr. Einstein been cloned 20 years ago, that clone might be more concerned with sports cars, the ERA, in vitro fertilized babies, and rock music than with mathematics and physics. I would speculate that he might be consumed with concern about polluted air, polluted water, and over-refined and artificially preserved food. He might well think that mathematics and physics are not relevant (he might even fail to consider relevant or not relevant to what?). And so the argument goes. A cloned dictator could be a clerk-typist and a cloned assassinated president might become a professor. I know of no competent psychologist or behavioral biologist who could or would predict anything about what the expression of an individual's genes as a new person would be 20 years hence.

CLONING AND COMPASSION IN THE MEDICAL ARTS

It has been suggested that human cloning can serve a humanitarian function. Consider, if you will, a marriage in which one of the individuals has a genetic defect. The partner without the defect could be cloned, the union would be blessed with one or more children, and the genetic defect eliminated.

But there may be practical problems involved in a woman carrying to term her own clone. It is possible for fetal cells to break loose during gestation and find their way to the mother's tissue. Pregnant women carry not only a baby but membranes and tissues of fetal origin. Some of the tissues are concerned with nutrition of the fetus and are known as "trophoblast" cells. Ordinarily, the immune system of the mother rejects the "foreign" trophoblastic tissue after birth. What happens to the mother who has fetal trophoblast cells spread throughout her body that are genetically identical to herself? Presumably, she would not have the capacity to reject the cells and a fragment of the fetus would become as malignant as any other cancer. Instead of having the pleasure of seeing whether her clone outdoes herself, she might die of cancer related to the pregnancy.

If a man or a woman is physically vigorous enough to consider having himself or herself cloned, but has a genetic defect that prevents reproduction, the defect is probably recessive. The examination of cells and fluid from within the pregnant uterus, amniocentesis, has proved useful for detecting subtle genetic differences. Given these two facts, I wish to make a modest proposal: Instead of resorting to cloning which may be even less available in the future than heart transplants are now, why not let the couple who fear genetic disease in their offspring utilize antenatal diagnosis. Even if both partners carry recessive genes for the defect, only one of four embryos will manifest the condition or have the potentialities for the condition. That unfortunate fetus, at most a 25% risk, could be diagnosed early in pregnancy and terminated by legal abortion. One may quibble about appropriate antenatal tests being not yet available for all genetic disease. Permit me to remind the reader that human cloning and mouse cloning, for that matter, is still well in the future, but examination of amniotic fluid and amniotic cells already permits in utero diagnosis of certain human diseases.

It might be a character-building exercise to adopt a child and forget cloning under the conditions described. If one parent has the strength to forgo biological parenthood because of possible genetic effects, I would suggest that the other parent emulate the first, do a socially constructive thing, adopt and enjoy parenthood without adding to the biological burden of an overpopulated planet.

Additionally, there is the unresolved problem of whether a mother is a mother to the clone of her husband. Is the father the father or a brother to his infant clone? And will love or resentment emerge when the asexually created twig realizes that he or she was created on a bench with a micromanipulator instead of as a consequence of sexual love? Personally and subjectively, I find that in a chaotic world with an uncertain future, both a mother *and* a father are reassuring. Humans are notoriously resistant to advances in the technology of parturition. Perhaps my apprehensiveness about the lack of a conventional parent may be in the same category as the fear voiced by some of painless childbirth. I admit that my reaction is emotional, not scientific.

I freely admit to love of parents and love of family. I also recognize that I would love any young individual reared in my home. So, why not love the cloned offspring of my mate? Conversely, why shouldn't my spouse have concern and affection for my cloned self?

The relationship would be novel. Perhaps learning, because

the genetic makeup is identical, would be enhanced. I suspect that whether or not a cloned individual communicated more readily with its genomic parent than it would with anyone else would depend upon individual circumstances. It seems to me just as likely that the clone would reject its single "parent" as it would be especially fond of that parent. I know of no way that one could predict whether rejection or acceptance would occur.

Even though an attempt has been made, I still think, as I said earlier, that I will not witness cloning of a human being. Although it might be compassionate to assist a childless couple to have a child through cloning, the relatively elaborate technology required for this procedure does not address any major medical problem. Is it moral to spend large sums of money for biomedical equipment and salaries for technical personnel to *reproduce* people when worldwide lack of people is no problem at all? When overpopulation is the problem?

Are there medical problems that could be addressed with the technique?

SPARE PARTS—A REASON FOR CLONING?

The acceptance of tissue transplants without rejection among members of an isogenic group is a possible benefit of the cloning technique. Rejection of grafted tissue among individuals within a clone does not occur, just as an identical twin does not reject surgically transplanted tissue obtained from its genetically identical sibling. Because of this, some writers have suggested that one or more clones be fabricated to serve as a source of spare parts for a nuclear donor. Mechanics refer to the dismantling of an intact engine for parts to repair another engine as "cannibalism"—it is my feeling that the term is particularly apt in the situation proposed. It seems to me that it might well be ego-shattering to find at age 21 that one was fabricated on the bench so that one's clonemate could have a spare kidney should it be needed. It could be proposed that if one wished to utilize human cloning for spare parts, one should not make intact whole people, but should manufacture specific spare parts and then only as they are needed.

Consider the situation of a president of the United States with an aging and infirm heart. Both Presidents Eisenhower and Johnson were afflicted with heart trouble. I suggest that it would be feasible to obtain a microscopic biopsy from an ailing chief executive, and, either directly or after in vitro culture, insert a nucleus from a cell of the biopsy into an enucleated egg from an appropriate ovulated woman. If the cell-biology theory relating to the genetic equivalence of adult nuclei is correct, it would not matter from what type of tissue the biopsy was obtained. Thus, as the theory permits, skin, bone marrow, or liver would be equally acceptable as a source of diploid nuclei for transplantation.

But how is a heart obtained from the proposed cloning operation? Let me postulate that a heart could be produced in the following manner. The cloned president would be placed in the uterus of an appropriate surrogate mother, who would not be expected to carry the clone to term. Within the first month of embryonic development, only weeks after uterine implantation, and many months before "quickening," the heart is differentiated embryonically and autonomous beating begins. Appropriate organ culture techniques are not yet available, but may be in the not distant future. The organ rudiment, in this case a heart, would be removed from the cloned embryo and put in culture. The remainder of the embryo, not yet "alive" by some people's standards, would be discarded. Growth is rapid in organ culture and there is no need to predict that an inordinate growth period would be needed. A fertilized egg grows from a speck to 6 to 10 pounds in nine months. A heart of appropriate size for an adult might be grown in half that time. The chief executive who provided the biopsy would then become the recipient of the cloned heart.

The proposed transplant system stands in stark contrast to the macabre wait for the sudden demise of a suitable donor. After the dead donor's heart is transplanted, the recipient must evermore be monitored for possible rejection of the organ that sustains his or her life. How much better it would be to implant a new heart, genetically identical to the first, but more youthful by a half dozen decades. Critics will say, and accurately, that the isogenic implanted heart, *because* of its very identicalness to the former heart, would become prey to the same disease that afflicted the first. True. However, if it took six decades for the first heart to succumb, even if the second lasted only half that time, it would serve a life-prolonging function. Further, it could be that the second heart would benefit from increased knowledge of those insults that make American hearts so vulnerable to disease. Perhaps the chief executive with his or her new heart could be coaxed into not smoking or cajoled into consuming fewer fatty steaks. It seems to me that if one wants to prognosticate a suitable use that is potentially scientifically feasible for the cloning procedure, one may very well wish to consider the tailor-making of organs for the maintenance of already existing life.

The proposal for spare parts, like any other activity that requires immense technical skill and money, would never be available for all. Who would decide who gets a cloned heart? Who would be the recipient? There are, of course, many precedents for the inequitable distribution of medical care. Recently, the late Duke of Windsor became the recipient of a piece of artificial aorta in Houston. Surely his life was extended. Would we deny the duke his extra years because others were unable to seek competent medical diagnosis and superb surgical assistance to be equally served?

Spare parts—how simple. Not resolved, however, is the problem of risk to the surrogate mother, the risk that accompanies any pregnancy. Also not solved is how to dispose of the fragment of embryo "not alive" by some people's standards. I am not a lawyer and so I have personal difficulties with legal definitions that attempt to define when life begins. As a biologist, I think of the continuity of life—and I believe that the "lifeless" fragment that provided a heart rudiment is, in fact, a part of the continuity of life—and the surgeon who would dissect from it a heart would be terminating the life of an unborn human. I would not do that.

Whole books have been written about the ethics of abortion. It is not my purpose to consider the morality of such a sensitive issue here. Rather, it is my purpose to illustrate how human cloning might be medically useful. If human cloning becomes a reality, then society through elected representatives must pass enabling legislation (to protect the physician) or enact laws that proscribe the procedure (to prevent abortion for cloning or other surgical acts considered undesirable). I suggest that these legislative acts will be promulgated only when the surgical skills become available.

I also suggest that the issues are far more complex than is apparent at first sight. If we had the resources to clone a human to save a significant life, would abortion of the clone (to provide an embryonic heart) in fact be abortion? I ask this question because is not the significant act of a conventional abortion the termination of a life? The clone would be a cellular extension of an already existing life and termination of the embryonic clone does not terminate the individual who donated the diploid nucleus. Further, termination of the cloned embryonic life in the hypothetical situation would have the potentiality of *extending* the life of the donor of the diploid nucleus. Thus, would an abortion be an abortion in the special sense when its only significant feature would be life extension? I do not know the answers to these questions, and I doubt if many of the writers who suggest that cloning be used to provide spare parts do either.

There are other ethical considerations regarding cloning.

MORE ON ETHICS AND MORALS

If human cloning does happen, there is a real potential for developmental abnormalities. There is no reason to believe that manipulation of human eggs and nuclei would be anymore free of technical error and mishap than manipulation of amphibian eggs and nuclei. Is this a reason *not* to clone? I believe not. Abnormal offspring are a real hazard of conventional sexual reproduction. There is no guarantee that a fetus will be entirely normal. Although the statistical chance of birth defects may be calculated at about 7%, parents never know in advance what problems may occur during any particular pregnancy. Nature is imperfect and so is the potential synthetic human produced by cloning. Should risk of imperfection be a deterrent in cloning anymore than it is a deterrent in sexual reproduction? My answer to this is "No," provided there is good reason to clone in the first place. I have already suggested that I believe *reproduction* by cloning is not desirable—not all agree with me.

There are other questions. Since cloning will allow for close inspection of the operated egg, who will have the responsibility of terminating the life of a transplant egg that appears to be developing abnormally? I suspect that an abnormally cleaving egg will die before it becomes an abnormal baby—this is generally true (with exceptions) of amphibian eggs. Since abortion of a conventionally conceived fetus is legal, I see no reason why a grossly abnormal egg should not be aborted in vitro. But is loss of an embryo before implantation in the uterus truly abortion? Again, I profess ignorance.

There are, of course, other problems. Who will be the surrogate mother, that is, who will provide a foster uterine environment for a nuclear transplant human? One way of dealing with that question is to avert it and propose that an artificial placenta be used for the nine months of fetal development. It is my judgment that we are far more advanced in the technology of ovulation and micromanipulation of nuclei than we are in the technology of devising an artificial placenta. Thus it may well be that the decision to clone humans will be made long before extracorporeal gestation is possible, which would liberate women from the burden of pregnancy.

Liberation from the burden (and joy?) of pregnancy calls to mind how, until very recently, artificial formulas were thought to provide new freedom to young mothers. Although it is true that bottles provide nourishment to the baby during the mother's absence, some thoughtful individuals now question if laboratory-devised food is as good for the newborn as is mother's milk, and they point out that both baby and mother seem to do better with the warm human contact of baby's lips to mother's breast. These comments concerning the technology of baby feeding may reveal some of my skepticism about the technology of extracorporeal pregnancy.

If human cloning is resorted to for whatever reason, some women will probably volunteer to be surrogate mothers and be chosen as such. Is this morally undesirable? I do not know. One reason that it may be unacceptable is that it involves a risk to the life of a mother who is not even carrying her own child. I find risk repugnant but our society appears not to. Risk to the lives of certain people has always been acceptable to our western civilization and culture. Until recently we drafted young men into a combat force against their will. We expect asbestos workers and coal miners to expose themselves to high risk. We tolerate cigarettes. So why proscribe the actions of an individual who voluntarily offers her uterus as a home to a cloned human? I suspect that labor unions will have more to say about who and who does not offer her womb for wages than do the commentators on the moral scene.

SOULS—FRAGMENTED OR INTACT?

A biologist treads on quicksand when he/she attempts to make pronouncements concerning the soul. Although it may seem difficult for the layperson to think of cloned humans as having a soul apart from the nuclear donor, I believe he/she would have as much soul as does each twin of an identical twin pair or each triplet of a triplet trio when descended from a common fertilized egg. For monozygotic twins or triplets are as much clones as the clones that would be produced by humans at the laboratory bench. I would never argue that an identical twin has less soul than its ordinary sibling—and, in precisely the same way, a cloned human would have no less soul than does a human who develops as a result of a sexual act.

I like to think of the biological significance of separated blastomeres and cloning. There is another procedure that is quite opposite. This procedure involves bringing together two early embryos to grow and develop as one. The fusion of two embryos to develop as one normal embryo is as interesting in a developmental sense as is cloning. It probably has happened spontaneously in humans—just as occasionally a cleaving human egg splits and forms twins. Only in this case, two fertilized eggs fuse to form one individual. Does that one individual have two souls? I doubt it. And, if a cloned individual were ever produced, I doubt if it would have less soul than anyone else.

HAZARDS

Today we know more about hazard than humans have ever known before. We know what kills. Lack of exercise and too much saturated fat in the diet probably contribute to cardiovascular disease. Smoking probably kills more individuals than any other environmental insult, taking its toll through cardiovascular disease and cancer. The United States and other nations have thermonuclear instruments of death ready for instant deployment, making thermonuclear conflict an everpresent hazard.

I do not want to clone a human. I know of only one who has tried. I know of no one with the capability to clone a human. But what if a human were cloned? Would he or she be a hazard and would the act of cloning be a hazard? I doubt it. The cloned human would be a human just as the cloned frog is a frog. A human is not a hazard per se, although the acts of certain humans are hazardous. The cloned human might not be a hazard, but would the cloner be one? I doubt it. He or she would have an unusual skill—which would probably have little marketable value. If I had to limit myself to one skill, I would far rather be a machinist than a human cloner. There are many industries that make many useful products that will always have to employ machinists for the benefit of humans. I know of no industry now or in the future that will need the services of a human cloner.

Epilogue

Thomas Robert Malthus wrote in the early nineteenth century that the population of humans tends to exceed available resources. Charles Robert Darwin suggested just after the middle of the nineteenth century that the reproductive potential of a species is great, and he believed that the generation of enormous numbers of individuals played a role in evolution. Perhaps as Malthus and Darwin would have predicted, the population of the world in the late twentieth century is enormous. Crowding, pollution, and shortage of adequate food are serious problems. How, then, would it benefit mankind were humans to be cloned at the workbench of the cell biologist?

Earlier I wrote of genetic diversity. I said that we cherish uniqueness among our children and our friends. Uniqueness is at least partly the result of sexual reproduction. I cannot imagine why any reasonable person would want to minimize, through asexual reproduction, the heterogeneity that is essential to the richness of human lives.

Money is spent for biomedical research in the expectation that results of this research will enhance human health. What disease is healed by the production of genetic replicates? Cloning research continues because there are important biomedical problems that need solutions. Cloning is one of many experimental procedures that provide information that may help in solving problems such as cancer and aging. That is the why of cloning. And to assist people in understanding that "why"—i.e., the rationale of cloning—I have written this book.

Suggested Reading

Suggested Reading

Beauchamp, T. L., and L. Walters (eds.). *Contemporary Issues in Bioethics.* Encino and Belmont, California: Dickenson Publishing Co., 1978.

Braun, A. The reversal of tumor growth. *Scientific American* 213:75-83, 1965.

Braun, A. *The Story of Cancer.* Reading, Mass.: Addison-Wesley, 1977.

de Condolle, A. *Origin of Cultivated Plants.* 2nd ed. 1886. Reprinted New York: Hafner Publishing Co., 1959.

de Fonbrune, Pierre. *Technique de Micromanipulation.* Paris: Masson et Cie., 1949.

Developments in Cell Biology and Genetics. Hearing before the Subcommittee on Health and the Environment. Serial No. 95-105. Washington, D.C.: U.S. Government Printing Office, 1978.

Edwards, R. G., and R. E. Fowler. Human embryos in the laboratory. *Scientific American* 223:44-54, 1970.

Eisenberg, L. The outcome as cause: Predestination and human cloning. *The Journal of Medicine and Philosophy* 1:318-31, 1976.

Fletcher, J. *The Ethics of Genetic Control: Ending Reproductive Roulette.* Garden City, New York: Anchor Press/Doubleday, 1974.

Gurdon, J. B. *The Control of Gene Expression in Animal Development.* Oxford: Clarendon Press, 1974.

Hayflick, L. The biology of aging. *Natural History* 86:22-30, 1977.

McKinnell, R. G. *Cloning: Nuclear Transplantation in Amphibia.* Minneapolis, Minnesota: University of Minnesota Press, 1978.

Morgan, T. H. *Experimental Embryology.* New York: Columbia University Press, 1927.

Oppenheimer, J. M. *Essays in the History of Embryology and Biology.* Cambridge, Massachusetts: M.I.T. Press, 1967.

Pierce, G. B., R. Snikes, and L. M. Fink. *Cancer: A Problem of Developmental Biology*. Englewood Cliffs, N.J.: Prentice-Hall, 1978.

Ramsey, P. *Fabricated Man: The Ethics of Genetic Control*. New Haven: Yale University Press, 1970.

Ramsey, P. *The Ethics of Fetal Research*. New Haven: Yale University Press, 1975.

Ramsey, P. *Ethics at the Edges of Life: Medical and Legal Intersections*. New Haven: Yale University Press, 1978.

Spemann, H. *Embryonic Development and Induction*. 1938. Reprinted. New York: Hafner Publishing Co., 1962.

Steward, F. C. Totipotency, variation and clonal development of cultured cells. *Endeavor* 29:117-24, 1970.

Volpe, E. P. Embryonic tissue transplantation incompatibility in an amphibian. *American Scientist* 60:220-28, 1972.

Willier, B. H., and J. M. Oppenheimer. *Foundations of Experimental Embryology*. Englewood Cliffs, N.J.: Prentice-Hall, 1964.

ADDITIONAL PUBLISHED SOURCES OF ILLUSTRATIONS

DiBerardino, M. A., T. J. King, and R. G. McKinnell. Chromosome studies of a frog renal adenocarcinoma line carried by serial intraocular transplantation. *Journal of the National Cancer Institute* 31:769-89, 1963.

McClendon, J. F. The segmentation of eggs of *Asterias forbesii* deprived of chromatin. *Archiv für Entwicklungsmechanik der Organismen* 26: 662-68, 1908.

McKinnell, R. G. Intraspecific nuclear transplantation in frogs. *Journal of Heredity* 53:199-207, 1962.

McKinnell, R. G. Expression of the Kandiyohi gene in triploid frogs produced by nuclear transplantation. *Genetics* 49:895-903, 1964.

McKinnell, R. G., and K. S. Tweedell. Induction of renal tumors in triploid leopard frogs. *Journal of the National Cancer Institute* 44:1161-66, 1970.

Suggested Reading

Roux, W. Beiträge zur Entwickelungsmechanik des Embryo. Ueber die Künstliche Hervorbringung halber Embryonen durch Zerstörung einer der beiden ersten Furchungskugeln, sowie über die Nachentwickelung (Postgeneration) der Fehlenden Körperhälfte. *Virchows Arch. Pathol. Anat. Physiol.* 114:113-53, 1888. Resultate 289-91. English translation by H. Laufer. In B. H. Willier and J. M. Oppenheimer (eds.), *Foundations of Experimental Embryology*. Englewood Cliffs, N.J.: Prentice-Hall, pp. 2-37.

Volpe, E. P., and R. G. McKinnell. Successful tissue transplantation in frogs produced by nuclear transplantation. *Journal of Heredity* 57: 167-74, 1966.

Index

Index

(Page numbers for illustrations are in boldface)

Index

Index

Index

Index

Index